Being Evidence Based in Library and Information Practice

Being Evidence Based in Library and Information Practice

Editor

Dominika Dechiel

Being Evidence Based in Library and Information Practice

Edited by **Dominika Dechiel**

ISBN: 978-1-68117-252-1
Library of Congress Control Number: 2016934795

© 2017 by
SCITUS Academics LLC,
www.scitusacademics.com
Box No. 4766, 616 Corporate Way,
Suite 2, Valley Cottage,
NY 10989

Notice

Preface

Library and Information Science is a profession that is full of people passionate about making a positive change in the world & they tend to be wildly happy about what they do. The joint term is associated with schools of library and information science (abbreviated to "SLIS"). In the last part of 1960s, schools of librarianship, which generally developed from professional training programs (not academic disciplines) to university institutions during the second half of the 20th century, began to add the term "information science" to their names. Evidence-based library and information practice (EBLIP) or evidence-based librarianship (EBL) is the application of the interdisciplinary approach known as evidence-based practice (EBP) to problems in the field of library and information science (LIS). This means that all practical decisions made within LIS should be based on research studies and that these research studies are selected and interpreted according to some specific norms characteristic for EBP. Typically such norms disregard theoretical studies and qualitative studies and consider quantitative studies according to a narrow set of criteria of what counts as evidence. If such a narrow set of methodological criteria are not applied, it is better instead to speak of research based library and information practice. This book, Being Evidence Based in Library and Information Practice provide evidence summaries an opportunity for librarians and other information professionals on a wide number of topics that may contribute to decision making in professional practice.

Table of Contents

1

CHAPTER

AN EMPIRICAL REVIEW OF LIBRARY DISCOVERY TOOLS

Xi Shi*, Sarah Levy

SUNY Rockland Community College, New York, USA

ABSTRACT

The Internet search concept has fostered an expectation that all users need to do is to feed relevant terms to a search engine to describe a topic or to ask a question, and click "search". The search engine is then expected to return a list of possible relevant and useful results for users to choose from. Based on this search concept, library system developers have been developing and constructing software programs for library databases that manage scholarly information. These software programs are known as library discovery tools, or web-scale discovery (WSD) tools. In this article, the term "library discovery tools" is used when discussing search engines designed for the libraries, WSD is included. Library discovery tools are intended for intelligent searches for educational or research purposes. This article provides a practical analysis of available library discovery tools in context of the present-day explosion of available open search engines on the Internet. The focuses of our analysis include how discovery tools are expected to manage library collections, provide access to scholarly information content, as well as other factors, such as budgetary considerations, when choosing or adding a discovery tool for a library.

KEYWORDS

Information Search Engines, Library Discovery Tools, Information User Satisfaction, Information System Success

1. INTRODUCTION

Although the concept of library discovery tools is no longer novel to many, including librarians, the understanding of this tool is far from inclusive, nor is it conclusive as to what they do, how they do it, and the cost of delivering what is expected.

As both authors are librarians at a community college library, for the purposes of this article, we discuss library information management systems, library information search tools, particularly discovery tools, as IS products and services. Our library users, mostly our community college students, are observed and discussed as a specific sub-group of computing end users, or computing consumers. A library discovery tool is defined in this paper as a search engine that builds on unified indexes of licensed scholarly information, searches across multiple library databases provided by different vendors, and may be customized for its size, range and comprehensiveness of data inclusion for targeted solutions. Deliberations are presented on what search tools in academic libraries are most appreciated by college students and faculty, highlighting the differences between library search tools, including discovery tools, and commercial search engines.

Currently, the Internet is exploding with available free and fee based web-scale search tools for scholarly information, and many more are getting on Internet daily, of which some are commercial, such as Google Scholar, and others are sponsored by academies, such as Cite Seer X. Our analysis only includes those that are being commonly recognized and habitually utilized by college students for their comprehensiveness of academic information content coverage, and satisfying return of search results for discussion and examination.

A practical review is offered on how search tools are evaluated by users, and which tools are often identified and selected by specific clientele as appropriate for satisfying their information needs. First, a comparative analysis of commercial search engines versus library search tools is provided. Search experiences are discussed in order to describe characteristics of the various search engines, thereby identifying what information users expect from these search engines, and the factors users value when selecting one of many search engines for their information needs. Second, the most popular library databases and search tools are evaluated. The strengths and weaknesses are discussed to understand the practical functionality of each tool, and to facilitate the formation of realistic expectations of the currently available library discovery tools. Finally, financial cost-benefit considerations are reviewed as an additional factor to be weighed when library databases or search tools are being investigated.

2. COMMERCIAL SEARCH ENGINES

2.1. Google and Its Impact on the Library Market

Google, like it or not, has "created a model that librarians, as information providers, must meet head on." (Luther, 2003) [1] . It is in the best interest of librarians and library IS developers to understand Google, what it does, how it progressed to today's performance level and to learn from it, rather than distancing ourselves and our information users from it.

Most academic librarians do not, and generally speaking librarians should not, discourage students from using Google, especially Google Scholar, as Google Scholar never fails to provide relevant information for any set of search terms. Very often, Google Scholar includes full text information offered by educational institutions at no cost to the users. In some instances, the search will direct the users to their home library if the full text is found to be available via their library subscriptions. Even if the full text is not offered from the initial Google Scholar search, once the bibliographic information is provided, a targeted search by author(s) or title can be followed up in library databases rather easily.

Content retrieval from a Google search has no doubt been recognized by most educated users. Google's mega-data search algorithms, indexing preferences, ranking and displaying techniques have been observed, and its capacity of providing relevant contents, which is often described by many librarians as "good enough content" (Luther, 2003) [1] has been recognized and accepted. This is one reason that many library IS developers are imitating what Google does by creating

"Google like" search tools to manage library information resources. However, one information retrieval characteristic presented by Google that has not received as much attention is that the relevant information content from a Google search is not always displayed directly or instantly. This statement may appear to be conflicting with many existing findings in the literature discussing how much information content Google does provide. We noticed Google often functions as a pointer, when it does not present or display full text information content. It does so sometimes even when it does deliver full text information content, so as purposefully to provide additional information from different sources. For example, when searching in Google for the term "define plagiarism", a seven digit result is indicated. The first entry listed on the result page is a definition of plagiarism provided by Google. Following this full text definition are links that point to definitions offered by other sources, e.g., dictionary.com, Wikipedia, and academic dictionaries, such as Oxford Dictionaries; as well as definitions provided by many educational institutions, e.g., Princeton University.

Many studies have reported and confirmed that academic library users use Google, love Google and complete their assignments on Google (Al-Maskari & Sanderson, M. 2011) [2] ; (Gross & Sheridan, 2011) [3] ; (Luther, 2003) [1] ; (Thompson, Obrig & Abate, 2013) [4]). We believe this user behavior is likely influenced by user's expectations. However, how the user's decision is made to turn to Google and return to Google still needs comprehensive investigations.

2.2. Search Engine Success Determinants

In the IS field, both researchers and practitioners, such as software developers, believe that user satisfaction is an indication of successful IS product design (Au, 2008) [5] ; (Bolton & Drew, 1991) [6] ; (Heinbokel, et al. 1996) [7] ; (Hsieh et al. 2012) [8] . This same belief is shared in library science (Condit Fagan et al. 2012) [9] ; (Gross & Sheridan, 2011) [3]). It is well researched that satisfied information users tend to continue with the system or programs they have experienced, and dissatisfied users will most likely look for different tools for the task at hand, or return to the tools with which they had prior positive experience. It is well documented that the influence of user satisfaction is closely associated with the selection of library IS products.

According to satisfaction researchers, consumer satisfaction is heavily influenced by consumer's expectations of a product or service being consumed. Therefore, many studies define satisfaction formation process as a comparison process in which consumers compare their pre-purchase expectations with the product or services after the purchase (Bhattacherjee, 2001) [10] ; (Churchill & Surprenant, 1982) [11] ; (Shi et al., 2004) [12] ; (Spreng et al., 1993) [13] , (Spreng et al., 1996) [14] . The most renowned theory that has been applied to explain information user satisfaction in both IS and library fields is disconfirmation of expectations. This theory proposes that consumer satisfaction is a result of a comparison of pre and post purchasing experiences. Before the purchase, consumers have expectations of what they are buying. During the purchasing process, consumers evaluate and select the products and services according to their pre-purchase expectations. After the purchase is made, consumers compare what they have purchased with their pre-purchase expectations. If the products or services exceed consumers' expectations, satisfaction occurs. If the products or services fall short of consumers' expectations, dissatisfaction occurs. In other words, disconfirmation is defined as the discrepancy between the consumers' expectations and the actual purchase experiences. Outcomes of the consumption that are better than expected lead to positive disconfirmation. Outcomes of the consumption that are worse than expected lead to negative disconfirmation.

As for information users, specifically library users who are consumers of IS products and services, satisfaction of search results is heavily influenced by their expectations. On one hand, it has been recognized that Google has changed information seekers' expectations, and defined and redefined the search experiences of those who are seeking information, scholarly and/or other kinds. On the other hand, library IS developers as well as librarians, have created very high users' expectations for library databases and discovery tools for scholarly information searching, promising discovery tools will do equally well, if not better than Google.

For the purpose of a comparison analysis, one search example is offered. Searching Google for the article Cognitive Test Anxiety and Academic Performance, by J.C. Cassady & R.E. Johnson in Contemporary Educational Psychology, the first listing shows the exact article with a brief description of this article. Clicking on the listing opens the article in Science Direct, and the full text can be retrieved in PDF. In addition, attached to the listing, users can easily find "Cited by" information, formatting guides to "Cite" in APA, MLA and Chicago. Or, user may choose "Save" the listing into "My library" for later review. Please note, Science Direct is one of our library's subscription databases, although no effort was put in from the library to have this article listed, and linked to the database by Google.

The same search is then conducted in Ebsco Discovery Service (EDS), which our library subscribes to. Searching by exact title, nine records are displayed. One of the nine records shows the exact title, authors and the publication information. However, the user is being pointed to Scopus, a database that our library does not subscribe to. If the search is revised to exact title and author(s), limiting to full text only, two records are listed. The record for the article from Scopus still shows, although no full text is available to our library users. The other record shows the title as "Regular Article: Cognitive Test Anxiety and Academic Performance", with a link indicating "View record from Science Direct", which will retrieve the full text article.

The above sample search demonstrates that the Google search probably exceeded the user's expectation, in a sense that the full text is quickly displayed in the search results, and can be retrieved with no additional effort from the user, and at no cost to the user. In addition, citation guides are handily provided for college students, and "Cited by" information offers immediate access to other research articles mostly in the same field where the original article was cited and discussed. On the other hand, the library discovery search tools, such as EDS, requires intensive customization from the library end, including constructing a "profile" to include all subscription databases, journals and other print and electronic materials. Regular updates of all holdings is required, including the loading of catalog MARC records if the library wishes their book collection to also be incorporated in the discovery search. Also, updating library collection information is the responsibility of the library, e.g., additions and deletions of journals and databases to the library collection. On the other hand, Google requires no effort from the user end to refer a search to a more appropriate database. All Google search features are integrated into one interface and any given search will be directed to the more appropriate search tools, which can also be activated at "More" and "Even more from Google". For example, if a search is detected as possibly being a scholarly article search, the search is automatically redirected to Google Scholar. Similarly, if a search is detected as likely being a book search, the search will go to Google Books. In addition, as Google announced in December 2009, it would begin to customize search results based on information gleaned from the user, indicating searches conducted by different individuals might receive different outcomes for their search results that are believed to better fit what those individuals have expected (Petter, DeLone & McLean, 2013) [15] .

3. USERS' PERCEPTIONS AND EVALUATIONS OF LIBRARY DISCOVERY TOOLS

In an information dynamic environment, consumers look for one stop shopping sites for their daily necessities, whether on the Internet or at local stores. Similarly, library users look for one-click search engines to fulfill their information needs.

The most widely recognized WSD services in the library market include Ebsco's Discovery Service, Ex Libris' Primo Central Index, Serial Solutions' Summon Service, and OCLC's WorldCat Local. Although these discovery tools are all web-based computer programs, the design of each was not based on the same conceptualization. The following discussion will explore some aspects that we believe are essential to understanding discovery tools in general, before attempting to select the most appropriate one for a specific library.

Ebsco's Discovery Services (EDS) is one of the discovery search tools highly recommended by librarians. Ebsco has won many academic contracts for the coming years. The State University of New York (SUNY), for example, signed a multi-year contract with EDS in 2013.

EDS obtains all licensed data from other suppliers. Wherever possible, EDS copies the bibliographic data and then includes it in Ebsco's metadata for processing and retrieval. By theory, this conceptualization should provide seamless searching across all information suppliers as well as direct retrieval without the use of pointers or linkers, since the data now resides with Ebsco. However, if any information or data is not included in Ebsco's central index, for example, if a supplier does not agree to supply their data, the discovery service will either omit the information from the search or it will need to rely on a linker to retrieve the full text. Another design conceptualization depends on common programming standards, which are also labeled in the field of library science as "discovery layers" (Hoeppner, 2012) [16] . Summon is an example of this design approach. Layers apply when indexing across different platforms used by different data suppliers for data retrieval. Although the interface, display, and searching process for this design approach may present as a single application to the users, it is in fact only an improved federated search adjusted by adding more layers to include more data access, under the condition that all suppliers comply with the same standards.

The following provides a description of the authors' evaluations based on our own trials and experiences at our library.

EDS, a rather new product by Ebsco in the library market, is developed and marketed as a "Google like" search engine. EDS is supposed to return information search results from all library resources across different platforms that are provided by different vendors. After the users enter keywords or subject terms for a topic, EDS would then perform a smart search to lead the users to results as well as offering suggested adjustments to the original search, such as narrowing or limiting the search. It may also provide recommendations for further exploration, e.g., providing different search terms, or searching in different areas. In addition, it can provide directions and assistance for further steps in the research endeavor. One example of such assistance is the citation guides that include most academically accepted citation formats. Ebsco has a solid reputation for their comprehensive collection content, sophisticated search options, and responsive technical support. However, with information content from other vendors included, such as Science Direct, LexisNexis, etc., the discovery module does not appear to be designed to return search results on an equal ranking across the providers. Some articles might be missing from the EDS search results, but would be retrieved if a search is conducted in the native database where the articles reside. This phenomenon confirms the bias of information retrieval as reported in other studies that confirm that the bias of search engine coverage and the bias of information retrieval system does exist (Buttcher & Soboroff, 2007 [17] ; Vaughan & Thewall, 2004 [18]). Consequently, users must understand that the search results provided by a discovery tool from one specific vendor may not be content neutral, whether it is due to the indexing technique or the classification standards. In other words, the Ebsco Discovery Service user should expect to be pointed to Ebsco information first and foremost.

From our experience, help from librarians often enhances the EDS search outcomes. As a matter of fact the librarians' assistance, which might include subject phrasing choices, searching technique guidance, is found most effective as library users appreciate librarians comprehensive knowledge not only regarding the subject matter of the information content, but also the performance of library subscription databases. Unlike Google searching, librarians are expected to be familiar with the content specialty of each database, search functionality of different vendors, as well as being knowledgeable about their institution's academic curriculum. EDS, although a discovery tool, works better with librarians' intervention matching the available information to satisfy the users' needs, in our case mostly students' information needs to complete their academic assignments. For example, librarians would recommend information from Science Direct as an intelligent choice for scholarly information on "joint injury", but for a specific legal case involving "joint injury" the user would receive better results in LexisNexis.

To better appreciate the consequences of the design concept in understanding how library discovery tools might function differently from one another, and how they differ collectively from commercial search engines such as Google, the following offers a factual analysis from our observations as librarians. The discussion of application evaluations and the users' selection of search tools are presented from three perspectives. First, the content orientation of the provider; second, users' information needs, information literacy level and their perceived net benefits, and third, budgetary or resource-based considerations and their influence on selection of a search engine, or search tool.

3.1. Content Orientation of Search Engine Providers

First, we should acknowledge that every library database has its own content orientation, in much the same way that commercial search engines are constructed. Although Google can be viewed as a discovery tool, when specific information is being searched for, the Google user will be redirected to a more appropriate search engine. Sophisticated users may directly choose a most effective engine for their particular search. For example, when looking for the definition of a term, a user may go directly to a dictionary or encyclopedia, or they might try Google first, and then be redirected to a more appropriate site. Most current generation college students know to go directly to Amazon when seeking to purchase books, to Wikipedia for quick information on a subject, and to their college website for academic related information, such as registration dates, an academic calendar or course information.

Library databases or search tools also have their own specialties in content collections. Therefore, the database construction concept of each provider may be influenced and governed by the distinctive knowledge of their collections, and created to work most efficiently with their own collections. For example, Gale specializes in literature; Ebsco started their library business with journal and serial publications; and OCLC is known as a bibliographic utility. Understanding the specialties of each vendor is important because it lands users in a position to form realistic expectations. For example, if users recognize that Gale specializes in literature content, they may reconsider if the use of a discovery tool is most sensible to find literary criticism, even if the Gale content is included in the discovery module. Understanding that Ebsco began as a journal servicer, users can expect their serial data collection to be one of the most inclusive. Similarly, OCLC's WorldCat should provide superior functionality when searching catalogs for library specific holding information, resource sharing possibilities, etc. Periodical or journal information at the publication level provided by OCLC derives from MARC records entered or loaded from other sources. Consequently the information quality may not be predictable, because the host of such information content may or may not behave as described. Similarly, Ebsco may not be the best tool for searching library catalog information, especially union catalogs, since this information is loaded in MARC format from other sources, such as OCLC or a local library system, into the EDS module. Anytime data is introduced from an external source, uncertainty can be anticipated, and users may experience less reliable search outcomes, which will inevitably lead to less confidence in the searching process.

The following scenario offers two search examples to illustrate the impact of the selection of a search engine on a search outcome. Students in a social studies class are assigned to research the correlation between quality and the cost of higher education. The results from different search engines are listed in Table 1 below for analysis. The same search term "higher education cost and quality" was applied to all searches.

The data (retrieved July 2015) from Table 1 shows the search tools selected, the search strategy applied and the results retrieved from the initial search conducted when the assignment was received. Please note that of the 65,746 EDS results, 36,827 are full text articles and all Science Direct listings have full text. The Google Scholar search was not initiated by the user. Instead, the Google Scholar search results were pointed to by Google web.

We conducted this same search in EbscoHost with Academic Search Complete, Education Source and ERIC simultaneously selected. The search returns 4813 results and 1488 are displayed when limiting these results to full text scholarly journals. See Table 2 for details of this search.

After the authors' careful evaluation of the results from this search, we found that the EbscoHost search results displayed in Table 2 provided much more focused content than the same search performed in the Ebsco EDS, which, like most library databases, has the ability to suggest subject terms within the search context to narrow down the results. Most databases also build in the "smart search" functionality to suggest similar or broader terms for expanding a search.

Our students, as is the case for most undergraduates, prefer to spend the least possible time on receiving or retrieving a minimum amount of acceptable information (Gross & Sheridan, 2011) [3] ; (Hoy, 2012) [19] . This behavior will be further discussed in later sections of this paper. From the analysis of our sample searches, two indications as to why the selection of a more effective search tool is important warrant attention. One indication is that a wise selection of a search tool can result in speedy and satisfactory information retrieval, creating an immediate positive perception by users of the search tool for their information search process. Second, this positive perception will most likely be reinforced by repeating positive experiences when selecting an effective search tool. In turn, the positive experiences enhance users' learning practice of evaluating available search tools, and then selecting the best one that may satisfy their information needs.

The following sample searches are provided to further analyze the importance of understanding the impact of selection of an effective search tool. Searching in Google for "Wharton Puts First-Year MBA Courses Online for Free" published by Business Week, September 13, 2013, pulls up the full text article. The same search was conducted in Ebsco EDS, EbscoHost, ProQuest Central, and Gale. No result was found from any of the tested databases, nor was the user pointed to other site(s), although Business Week (also indexed as Bloomburg Business week and Businessweek.com) is covered by Ebsco Business Complete to which our library subscribes, and it is also included in our individual journal subscriptions. The retrieval failures of newspaper articles, including from major papers such as The New York Times and The Wall Street Journal, are often found from the library databases, including discovery tools. Excuses from the providers include the incompatibility of the content indexing from the dot com version and print paper, misinformation received from the newspaper(s), etc.

3.2. Information Users Needs and Perceived Net Benefits

To begin our discussion on users' information needs and the role those needs may play in information searching and retrieval, a brief literature review is provided.

Table 1. Search result comparison.

Search Engine	Search Method	Search Results
Ebsco EDS	Keyword(s)	65,746
Science Direct	Default	202,970
Google	Default	140,000,000
Google Scholar	Default	4,400,000

Table 2. Ebsco education research complete, academic search complete & ERIC.

Search Engine	Databases	Search Method	Search Results
		Keyword(s)	4813
EbscoHost	Education Research Complete Academic Search Complete Eric	Limit to full text only	3262
		Also limit to scholarly journals	1488

Many satisfaction researchers in the field of IS and marketing consider it necessary to include a "needs" construct when studying information searching and retrieval. Findings of the theory development and test results were reported (Oliver, 1995) [20] ; (Spreng et al., 1996) [21] . Needs theory can be traced way back to the 1940s, when Maslow (1943) developed the theory of a hierarchy of needs. The basic assumption of needs theories indicate that when deficiencies of need are realized and identified, people are motivated to take actions to correct or reduce those deficiencies to satisfy the need. Since human beings have different levels of needs according to Maslow, actions taken would indicate different behavior patterns in an effort to fulfill those needs. Two areas of concern are worth noting. First, many satisfaction researchers argue that needs should be one of the driving factors in understanding consumer behavior, and that needs is likely to have some effect on predicting or influencing consumer satisfaction. Therefore, the construct of needs should be included into the disconfirmation equation. In doing so, some researchers find that consumers may compare the product performance, such as IS programs or search engines, with their identified needs. Accordingly, if the product performance exceeds the innate needs, the consumer realizes the personal benefits, thus is satisfied. On the other hand, if the product performance does not fulfill the pre-identified needs, for example, the search failed to provide the sought after information, the consumer would perceive no net benefits, thus is dissatisfied (Petter et al., 2013) [15] ; (Sirgy, 1984) [22] ; (Spreng et al., 1996) [14] ; (Shi et al., 2004) [12] .

Researchers in the field of information science noted another perspective on the impact of needs on satisfaction. This perspective denotes that unlike other human needs, such as the need for food, water or shelter, etc., "what is required to satisfy an information need is often not known to the individual concerned" (Cole 2011, p. 1216) [23] . Information needs must be placed in context of the specific user's situation in order to be meaningful. Saracevic (2007) [24] stated that while most research on information needs concerns the user-centered concept, computer scientists when designing information retrieval systems often ignore the information-situation condition. Because the nature of information needs is intangible and non-specifiable, the information users may be engaged in a long process of activities as they search and then revise their searches to find the needed information. Therefore, no matter how long the process of information searching takes and how it progresses, e.g., an academic assignment or school project may in fact last weeks or months, there will be different types of information searches throughout that time while the information needs for this academic assignment remain constant. This information seeking and searching behavior can be witnessed in libraries with students looking for information to satisfy their assignment or research project. Very often students come in to the library without knowing or understanding what information they may need, or where to start searching for the information that may fulfill their assignments. Over the course of doing and completing their assignment, students learn to recognize and better understand their information needs, and the information needs become clearer and clearer as the information seeking activities continue. It is a learning process, as students conduct information searches trying and using different tools to get and review various search results. This learning process helps the students realize and clarify their information needs. In the meantime, their experience with searching activities and the applications of those tools in their searches, forms expectations of what they can find, which tools are effective, where they should start, and how to get the most needed information in the shortest time.

Additional theories on how users' needs should be integrated into IS design warrants further review. Users' information needs have been widely studied and recognized by researchers in many different fields. However, how the needs are understood and integrated into the IS design is approached differently. The concept of cognitive approach is attention worthy. The cognitive approach considers that it is the information specialists' knowledge that determines the successful design of an effective information retrieval system, and that knowledge is not based on the empirical studies of users (Hjorland, 2013) [25] . Examples of this approach include Apple computers and iPhones. Apple's success suggests that understanding what users may want or need as individuals, as well as socially and culturally, offers insight for their design. Apple's success demonstrates that even if a company does not interact directly with users, even if the design is not based on a review of the market, it can still be much more insightful about what people could want (Verganti, 2009) [26] . The library catalog as an IS retrieval system is a similar example. The most popular information organization schemes for library catalogs used by American libraries include Library of Congress and Dewey classification systems. Whether a catalog is an LC or Dewey based system, it is not likely that most college students will show interest in learning to understand the classification scheme. Most students simply need a tool to identify and locate the needed information. In this situation, librarians' expertise in information organization and retrieval would be most appreciated in helping students find needed information quickly and effectively from a library catalog, which is now much more inclusive than just book collections. In addition, librarians' help, including instructions on how to best utilize the available search tools, benefit not only those students who often prefer to spend the least possible time on receiving or retrieving a minimum amount of acceptable information, especially undergraduates, but also those serious users who demand and care to check the quality of the available information (Hjorland, 2013) [25] .

When discussing characteristics of information users, equity theory should also be reviewed. According to DeLone and McLean's IS Success Model (1992) [27] , (2003) [28]), the IS system success can be measured by system variables, including system quality, information quality, and service quality; and user variables, including system use and user satisfaction. The outcome of the Model is "net benefits" perceived by either individual users or organization/business. The basic concept of equity theory is that consumers' satisfaction with a product or service is determined by a cost-benefit analysis. The consumer feels satisfied if he/she believes the input for the transaction to get the product is adequately rewarded. The input may include the time sacrificed, money spent, etc. In other words, if the consumers believe the time was justly spent and the product was well worth the money, they feel satisfied. On the other hand, if the consumers do not perceive any benefits of the consumption, for example, if they believe the time was wasted and the product or service was not worth the money paid, they will be dissatisfied (Boddy & Paton, 2005) [29] ; (Staples et al., 2002) [30] ; (Woodroof & Kasper, 1998) [31] .

Physical effort and time expended can be seen as two major inputs when using a library for educational materials or using a library database to find scholarly information. According to the equity theory, the end results of the library users' efforts, whether or not the materials are available in the library or available online from library subscription databases, will be judged against how much time and effort was required in order to retrieve the information. For example, does the student need to make a physical trip to the library? Is the information available from the library databases? How easy is it to identify, locate and retrieve the information? Does the database supply the full text right away, or only bibliographic information is offered and further searching is needed for accessing full text? Bear in mind that library users always conduct cost-benefit analyses, knowingly or unknowingly. From our previous sample search, for example, if a newspaper article can be easily pulled up from Google by a title keyword search, why should students go to the library homepage, then select a discovery tool or familiarize themselves with the database from which this newspaper may be retrieved?

4. CONCLUSION

This paper is offered to promote further discussions of library discovery search tools, their development, assessment and selection. The authors believe that individual library experiences with an applicable discovery product and the understanding of how a discovery product should function, calls for further deliberations by and among librarians and professionals in the IT field. With this objective in mind, the authors present their experiences and evaluations of these search engines. Practical examples are analyzed within the context of database design and management frameworks. Expectations of how library discovery products should work to satisfy our information users' academic and research needs are reviewed. The examples provided in this paper are all drawn from actual information seeking activities at our library, a community college library. These searches include both popular Internet sites, such as

Google, and library subscription databases. The relevant theoretical models that guided these evaluations are reviewed where appropriate. We believe the current market for discovery products is too assertive to expect a sensible and comprehensive decision from libraries. Librarians may feel compelled to provide their users with a discovery tool, whether or not these products are well-designed and mature enough to commit to, especially at their currently marketed price(s). If the determination is made that the discovery tools should be promoted, a careful evaluation of all products is recommended to ensure the selection of the best one to serve the targeted user group of a specific library. The library should then define within their budget limitations what they consider to be a reasonable expenditure for the selected product. In addition, we encourage librarians to investigate other models beyond the available discovery products on the library market. One suggestion includes an exploration of partnering with other unconventional contractors, for example, Google. Perhaps the library world could consider working with Google to develop a discovery tool designed for library tasks. This suggestion is derived from our experience with Google's design of an email system that has been employed by our college, and is now employed by many other academic institutions as well. Many of these institutions, ours included, trialed other mail protocols before adopting Google's Gmail. Considering this model, further exploration is recommended to develop more thoughtful and on-target library discovery search tools from the librarians' perspective with our users' needs in mind.

REFERENCES

1. Luther, J. (2003) Trumping Google: Metasearching's Promise. Library Journal, 128, 36-39.

2. Al-Maskari, A. and Sanderson, M. (2011) The Effect of User Characteristics on Search Effectiveness in Information Retrieval. Information Processing and Management, 47, 719-729.

3. Gross, J. and Sheridan, L. (2011) Web Scale Discovery: The User Experience. New Library World, 112, 236-247.

4. Thompson, J., Obrig, K. and Abate, L. (2013) Web-Scale Discovery in an Academic Health Sciences Library. Medical Reference Services Quarterly, 32, 26-41.

5. Au, N.N., Ngai, E.T. and Cheng, T.E. (2008) Extending the Understanding of End User Information Systems Satisfaction Formation: An Equitable Needs Fulfillment Model Approach. MIS Quarterly, 32, 43-66.

6. Bolton, R.N. and Drew, J.H. (1991) A Multistage Model of Customers' Assessments of Service Quality and Value. Journal of Consumer Research, 17, 375-384.

7. Heinbokel, T., Sonnentag, S., Frese, M. and Stolte, W. (1996) Don't Underestimate the Problems of User Centeredness in Software Development Projects—There Are Many! Behaviour& Information Technology, 15, 226-236.

8. Hsieh, J., Rai, A., Petter, S. and Ting, Z. (2012) Impact of User Satisfaction with Mandated CRM Use on Employee Service Quality. MIS Quarterly, 36, 1065-A3.

9. Condit Fagan, J., Mandernach, M., Nelson, C.S., Paulo, J.R. and Saunders, G. (2012) Usability Test Results for a Discovery Tool in an Academic Library. Information Technology & Libraries, 31, 83-112.

10. Bhattacherjee, A. (2001) Understanding Information Systems Continuance: An Expectation-Confirmation Model. MIS Quarterly, 25, 351-370.

11. Churchill Jr., G.A. and Surprenant, C. (1982) An Investigation into the Determinants of Customer Satisfaction. Journal of Marketing Research (JMR), 19, 491-504.

12. Shi, X., Holahan, P.J. and Jurkat, M. (2004) Satisfaction Formation Processes in Library Users: Understanding Multisource Effects. Journal of Academic Librarianship, 30, 122-131.

13. Spreng, R.A. and Olshavsky, R.W. (1993) A Desires Congruency Model of Consumer Satisfaction. Journal of the Academy Of Marketing Science, 21, 169-177.

14. Spreng, R.A., MacKenzie, S.B. and Olshavsky, R.W. (1996) A Reexamination of the Determinants of Consumer Satisfaction. Journal of Marketing, 60, 15-32.

15. Petter, S., De Lone, W. and McLean, E.R. (2013) Information Systems Success: The Quest for the Independent Variables. Journal of Management Information Systems, 29, 7-62.

16. Hoeppner, A. (2012) The Ins and Outs of Evaluating Web-Scale Discovery Services. Computers in Libraries, 32, 6-40.

17. Buttcher, S. and Soboroff, I. (2007) Reliable Information Retrieval Evaluation with Incomplete and Biased Judgments. Proceedings of the 30th Annual International ACM SIGIR Conference on Research and Development in Information Retrieval, Amsterdam, 23-27 July 2007, 63-70.

18. Vaughan, L. and Thelwall, M. (2004) Search Engine Coverage Bias: Evidence and Possible Causes. Information Processing & Management, 40, 693-707.

19. Hoy, M.B. (2012) An Introduction to Web-Scale Discovery Systems. Medical Reference Services Quarterly, 31, 323-329.

20. Oliver, R.L. (1995) Attribute Need Fulfillment in Product Usage Satisfaction. Psychology & Marketing, 12, 1-17.

21. Maslow, A.H. (1943) A Theory of Human Motivation. Psychological Review, 50, 370-396.

22. Sirgy, M. (1984) A Social Cognition Model of Consumer Satisfaction/Dissatisfaction: An Experiment. Psychology & Marketing, 1, 27-44.

23. Cole, C. (2011) A Theory of Information Need for Information Retrieval That Connects Information to Knowledge. Journal of the American Society for Information Science and Technology, 62, 1216-1231.

24. Saracevic, T. (2007) Relevance: A Review of the Literature and a Framework for Thinking on the Notion in Information Science. Part II: Nature and Manifestations of Relevance. Journal of the American Society for Information Science & Technology, 58, 1915-1933.

25. Hjørland, B. (2013) User-Based and Cognitive Approaches to Knowledge Organization: A Theoretical Analysis of the Research Literature. Knowledge Organization, 40, 11-27.

26. Venkatesh, V. and Goyal, S. (2010) Expectation Disconfirmation and Technology Adoption: Polynomial Modeling and Response Surface Analysis. MIS Quarterly, 34, 281-303.

27. DeLone, W.H. and McLean, E.R. (1992) Information Systems Success: The Quest for the Dependent Variable. Information Systems Research, 3, 60-95.

28. DeLone, W.H. and McLean, E.R. (2003) The DeLone and McLean Model of Information Systems Success: A Ten-Year Update. Journal of Management Information Systems, 19, 9-30.

29. Boddy, D. and Paton, R. (2005) Maintaining Alignment over the Long-Term: Lessons from the Evolution of an Electronic Point of Sale System. Journal of Information Technology (Palgrave Macmillan), 20, 141-151.

30. Staples, D., Wong, I. and Seddon, P.B. (2002) Having Expectation of Information Systems Benefits That Match Received Benefits: Does It Really Matter? Information & Management, 40, 115-131.

31. Woodroof, J.B. and Kasper, G.M. (1998) A Conceptual Development of Process and Outcome User Satisfaction. Information Resources Management Journal, 11, 37-43.

2

CHAPTER

A FULL TEXT RETRIEVAL SYSTEM IN A DIGITAL LIBRARY ENVIRONMENT

Kehinde Daniel Aruleba[1], Dipo Theophilus Akomolafe[2*], Babajide Afeni[3]

[1]Department of Mathematics & Computer Science, Elizade University, Ilara-Mokin, Nigeria
[2]Department of Mathematical Sciences, Ondo State University of Science and Technology, Okitipupa, Nigeria
[3]Department of Computer Science, Joseph Ayo Babalola University, Ikeji-Arakeji, Nigeria

ABSTRACT

The volume of information being created, generated and stored is huge. Without adequate knowledge of Information Retrieval (IR) methods, the retrieval process for information would be cumbersome and frustrating. Studies have further revealed that IR methods are essential in information centres (for example, Digital Library environment) for storage and retrieval of information. Therefore, with more than one billion people accessing the Internet, and millions of queries being issued on a daily basis, modern Web search engines are facing a problem of daunting scale. The main problem associated with the existing search engines is how to avoid irrelevant information retrieval and to retrieve the relevant ones. In this study, the existing system of library retrieval was studied. Problems associated with them were analyzed in order to address this problem. The concept of existing information retrieval models was studied, and the knowledge gained was used to design a digital library information retrieval system. It was successfully implemented using a real life data. The need for a continuous evaluation of the IR methods for effective and efficient full text retrieval system was recommended.

KEYWORDS

Full text, Information Retrieval, Library, Digital Library, Queries, Indexing, Catalogue

1. INTRODUCTION

For Centuries, libraries have been organizing reading materials on shelves for easy access. However, systematic methods that had been widely adopted for the organization of library materials and their recordings for use by readers came into being a little more than a century ago [1] . The term digital library is used to refer to a library where some or all of the holdings are available in electronic form, and the services of the library are also made available electronically-frequently over the Internet so that users can access them remotely [2] . The primary purpose of digital libraries is to enable searching of electronic collections distributed across networks, rather than merely creating electronic repositories from digitized physical materials.

An information retrieval (IR) system is designed to retrieve any documents or information required by the user community. It is primarily targeted to make the right information available to the right user at right time. IR is concerned with representing, searching, and manipulating large collections of electronic text data. IR is a discipline that deals with retrieval of unstructured data or partially structured data, especially textual documents, in response to a set of query or topic statement(s), which may itself be unstructured [3] . IR system does not inform i.e. change the knowledge of the user on the subject of his enquiry; it merely informs the user of the existence or non-existence and whereabouts of documents relating to the request.

Many problems are associated with the current system of IR and such can be seen from the inability of the system to process request timely and to present inadequate results among others. In view of these inadequacies, it is imperative to develop an IR system that will curtail these inadequacies.

2. RELATED WORK

The importance of IR keeps growing as the amount of digital information keeps expanding at an ever-increasing rate. Stored documents, photographs and contents of books, and billions of Web pages are useful only if they can be easily found when needed.

2.1. Information Retrieval Models

For effectively retrieving relevant documents by IR strategies, the documents are typically transformed into a suitable representation. Each retrieval strategy incorporates a specific model for its document representation purposes. According to [4] , the Boolean model is the first model of IR and probably also the most criticized model. Larson [5] shows that much of this criticism seems to be based on lack of knowledge about how to utilise its search possibilities. In this model, we can pose any query which is in the form of a Boolean expression of terms, that is, in which terms are combined with the operators AND, OR, and NOT. The model views each document as just a set of words.

The vector space model (VSM) represents documents and queries as vectors in multidimensional space, whose dimensions are the terms used to build an index to represent the documents. It is used in IR, indexing and relevancy rankings and can be used in evaluation of Web search engines. According to Shang [6] , the VSM procedure was divided into three stages. The first stage is the document indexing where content bearing terms are extracted from the document text. The second stage is the weighting of the indexed terms to enhance retrieval of document relevant to the user. The last stage ranks the document with respect to the query according to a similarity measure.

According to Gonzalez [7] , Language models (LM) for information retrieval are retrieval models (taken from the speech recognition field) that do not impose an explicit parametric form for the probability of relevance. Lafferty and Zhai [8] presented a formal connection between probabilistic and language models. The basic idea of the language modelling approach to IR is to assume that a query Q is generated by a probabilistic model of document D. In this context, the generative language models approach estimate £ᵢ is the probability of the query being generated by a document.

2.2. Query Types

There are many different ways of searching for information. Here we describe the most common ones according to Salerma [9] .

A normal query is any query that is not explicitly indicated by the user to be a specialized query. For queries containing only a single term, the desired semantics are clear: match all documents that contain the term. For multi-word queries, however, the desired semantics are not so clear. Some implementations treat it as an implicit Boolean query by inserting hidden AND operators between each search term.

Phrase queries are used to find documents that contain the given words in the given order. Usually phrase search is indicated by surrounding the sentence fragment in quotes in the query string. They are most useful for finding documents with common words used in a very specific way [9] . For example, if you do not remember the author of a paper, searching for it on the Internet as a phrase query will in all likelihood find it for you.

Boolean queries are queries where the search terms are connected to each other using the various operators available in Boolean logic, most common ones are AND, OR and NOT [10] . Usually parentheses can be used to group search terms. A simple example is software AND database, and a more complex one is software AND database AND data structure.

3. INFORMATION RETRIEVAL AND DIGITAL LIBRARIES

Libraries have been in existence since the beginning of writing and have served as a repository of the intellectual wealth of society. As such, libraries have always been concerned with storing and retrieving information in the media it is created on. As the quantities of information grew exponentially, libraries were forced to make maximum use of IR methods to facilitate the storage and retrieval process. Some of the IR methods used in digital libraries are described in the following:

Indexing: IR systems need an indexing mechanism for performing efficiently the retrieval process [7] . Indexing is the transformation from the received item to the searchable data structure. Building an index from a document collection involves several steps, from gathering and identifying the actual documents to generating the final data structures [11] .

Catalogue records: are short records that provide summary information about a library object. The word catalogue is applied to records that have a consistent structure, organized according to systematic rules. Library catalogues serve many functions, not only information retrieval. Some catalogues provide comprehensive bibliographic information that cannot be derived directly from the objects. This includes information about authors or the provenance of museum artefacts [12] . Descriptive Metadata: Many methods of information discovery do not search the actual objects in the collections, but work from descriptive metadata about the objects [12] . The metadata typically consists of a catalogue or indexing record, or an abstract, one record for each object. Usually it is stored separately from the objects that it describes, but sometimes it is embedded in the objects. Descriptive metadata is usually expressed as text, but can be used to describe information that is in formats other than text, such as images, sound recording, maps, computer programs, and other non-text materials, as well as for textual documents.

Library Digitization

According to Ian and David [13] , defined digitization as the process of taking traditional library materials that are in form of books and converting them to the electronic form where they can be stored and manipulated by a computer.

According to Alhaji [14] , there are three main reasons for digitization of a library system

1   To make the documents more accessible: This is to serve existing library users better, i.e. to allow users search the full text of documents or to allow users search from remote locations.
2   To preserve the documents: Allow user read older or unique documents without damage to the originals
3   To reuse the documents: Allowing conversion of documents into different formats.

4. METHODOLOGY

From the architecture of a FTRS described in Aruleba et al. [15] , a full text search retrieval system was designed. This section presents the modelling of the system. The modelling is in two parts, which are: Analysis and Design

The analysis of the existing information system in University of Ilorin library was extensively carried out by studying the existing environment. The result of the analysis is presented using the Use-case diagram shown in figure 1.

The result of the existing library information shows that the entire system is made of seven steps. The first step is where the potential library user is registered. The registration allows the user to become a registered and legally authorised user of the library. After the registration, the user is allowed to use the facilities provided by the library. After registration, the registered user is allowed to undertake the remaining steps that is registered user can check available books, read books, check document title, author, publisher, borrow book, return book and pay fine in case of late submission of book(s).

The proposed system in addition to the functionality of the existing system allows users to search, modify user details, and upload documents as shown with use-case in Figure 2.

From Figure 2, the proposed system is made up of eight distinct steps. Though the components are interwoven, each of them performs distinct functions but all work together as a system to process request timely.

4.1. Database Design

Database design mainly includes requirement analysis, concept structure design stage, the logic structure design stage, physical structure design stage, database implementation stage, database operation and maintenance stage, there are six steps altogether.

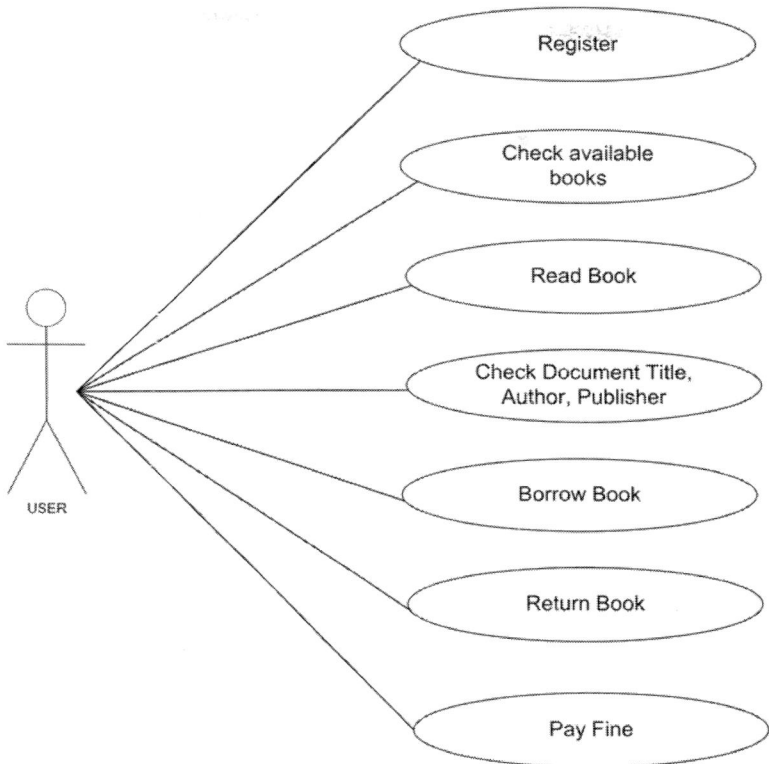

Figure 1. The result of analysis of UNILORIN library system.

From the analysis done, Table 1 was designed for the implementation of the proposed system.

4.2. User Interface Design

There are many factors that must be considered when designing the user interface of a software because the user must be able to interact with the system in a way that the system will understand whatever input given by the user. Therefore, the quality of the interface and software in general must pass the usability testing standard. Some usability factors, such as fit for use, ease of learning, task efficiency, ease to remember, subjective satisfaction and understand ability but all are put into consideration when designing the user interface (Figure 3).

The home page screen depicted in figure 4, contains four major modules which are the Search, Registration, Request and Login while the Admin module home page shown in figure 5, contains Sub-module which are view students, view staff, view books, create new book, view book request, create/view facilities, create/view department, logout. Each of them will lead you to its database when clicked and manipulated.

5. SYSTEM IMPLEMENTATION PHASE AND TESTING

This phase implements what have been discussed in thesection 4. The system was developed and implemented with PHP and MySql Technology.

Table 1. Generated database.

Table	Action
Admin	To authenticate the system user
Book request	To store information about requested books by the system users
Books	To store book details
Department	To store all the departments available
Faculty	To store all the faculties available
Staff	To store information of staffs using the system
Students	To store information of students using the system.

Figure 2. Use case diagram showing the proposed system.

Home Page Interface Implementation

The home page shown in figure 6 is the key aspect of the system, because it gives the basic user interface for the full text retrieval digital library. It comprises of: Search, Login, Registration and Request described as follows:

Search: This feature can be used by any user. This module provides a convenient book searching function, the user could search books based on a variety of conditions.

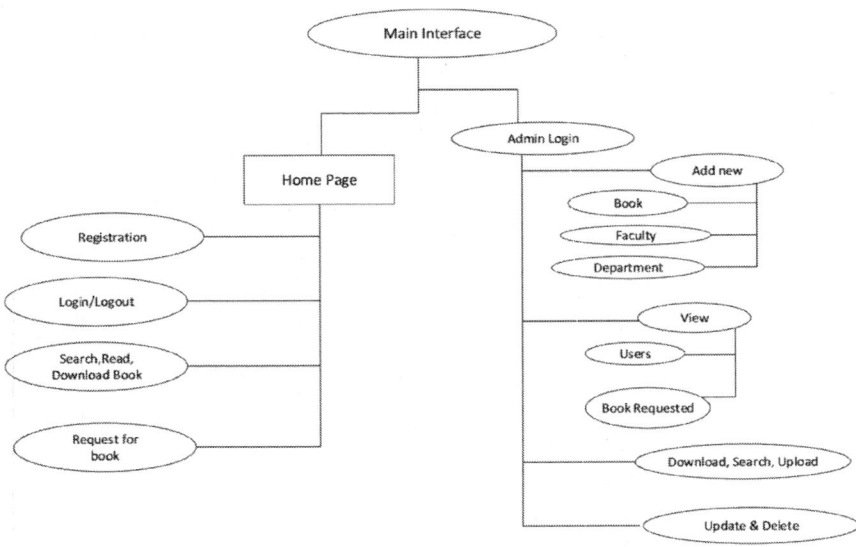

Figure 3. Showing the main interfaces of the system.

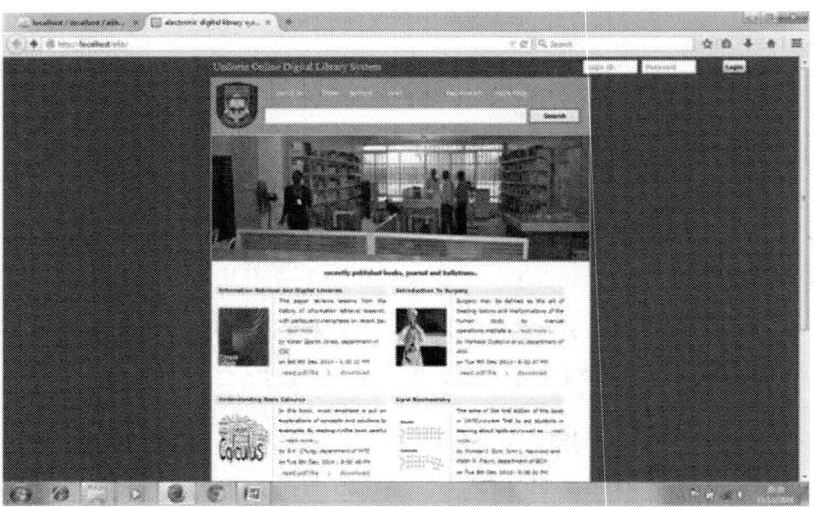

Figure 4. Home page design interface.

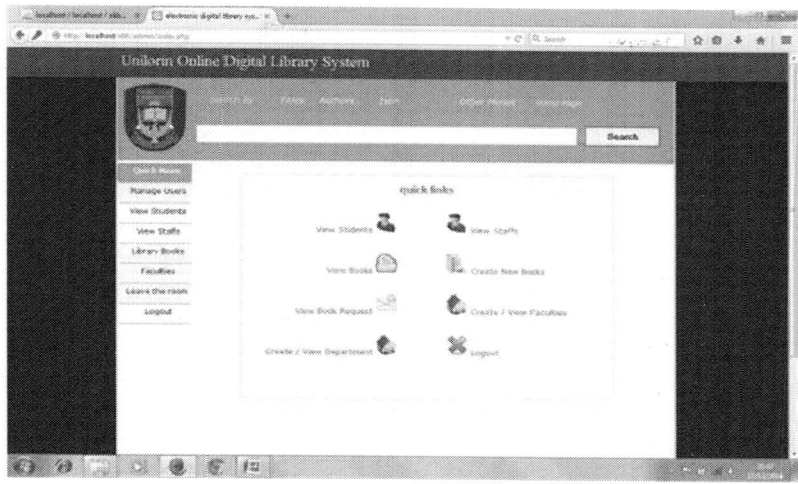

Figure 5. Admin home page design interface.

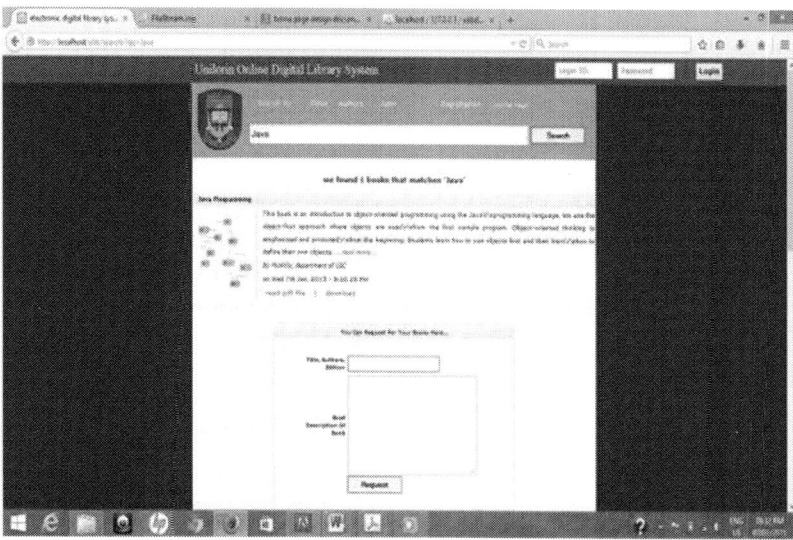

Figure 6. Home page implementation output.

Login: Every user who wants to use the system is authenticated by means of username and password. All entered parameters of the password are matched with information stored in the database, therefore only authenticated users can log on to the program with limited access.

If the login information is wrong, the user will be notified of login failure and would need to try again.

Registration: This involves registering new users. It contains registration form interface with entries like email address, last name, first name, password, password confirmation and sex.

Request: If a user can find the specific book needed, request can be made for such book.

6. CONCLUSIONS AND FUTURE WORK

This study has been able to develop and successfully implement an IR system that reduces the hurdles associated with present system of searching in Libraries. It is shown in the system that users searching full text are more likely to find relevant articles than searching only abstracts. This finding affirms the value of full text collections for text retrieval and provides a platform for aligning searching algorithms that take advantage of rapidly-growing digital archives.

Also, the following areas of the study can be improved upon in future studies to create a more robust IR:

- Increase in the size of the database, this will enable large data storage.

- Integrate advert plans for research materials the institution wants to be selling online.

- Acquire and publish video, audio and heavy graphic research materials.

REFERENCES

1. Onwuchekwa, E.O. and Jegede, O.R. (2011) Information Retrieval Methods in Libraries and Information Centers. An International Multidisciplinary Journal, 5, Serial No. 23

2. Rosenberg, D. (2005) Towards the Digital Library: Findings of an Investigation to Establish the Current Status of University Libraries in Africa. International Network for the Availability of Scientific Publications, Oxford.

3. Greengrass, E. (2002) Information Retrieval: A Survey by Ed Greengrass. Information Retrieval, 141-163.

4. Goker, A. and Davies, J. (2009) Information Retrieval: Searching in the 21st Century. Wiley, West Sussex.

5. arson, R.R. (2010) Information Retrieval Systems. In: Bates, M.J. and Maack, M.N., Eds., Encyclopedia of Library and Information Sciences, 3rd Edition, CRC Press, New York, IV, 2553-2563.

6. Li, L.Z. and Shang, Y. (2000) A New Statistical Method for Performance Evaluation of Search Engines. ICTAI.

7. González, R.B. (2008) Index Compression for Information Retrieval Systems. Unpublished PhD Thesis, University of A Coruna, A Coruna.

8. Lafferty, J. and Zhai, C. (2003) Probabilistic Relevance Models Based on Document and Query Generation. In: Croft, W.B. and Lafferty, J., Eds., Language Modelling for Information Retrieval, Kluwer, Pittsburgh, 11-56.

9. Salerma, O. (2006) Design of a Full Text Search Index for a Database Management System. MSc Dissertation, University of Helsinki, Helsinki.

10. Pazer, J.W. (2013) The Importance of the Boolean Search Query in Social Media Monitoring Tools. Dragon Search.

11. Manning, C.D., Raghavan, P. and Schutze, H. (2008) Introduction to Information Retrieval. Cambridge University Press, Cambridge.

12. Arms, W. (2002) Manuscript of Digital Libraries. MIT Press, Cambridge.

13. Written, L.H. and Brainbridge, D. (2003) How to Build a Digital Library. Morgan Kaufman Publishers, London.

14. Alhaji, I. (2005) Digitization of Library Resources and the Formation of Digital Libraries: A Practical Approach. University of Pretoria, Pretoria.

15. Aruleba, K.D., Aremu, D.R., Oriogun, P.K., Agbele, K.K. and Agho, A.O. (2015) Evaluation of Full Text Search Retrieval System. Nigeria Computer Society, 26, 154-159.

3

CHAPTER

USING THE WORLD WIDE WEB TO CONNECT RESEARCH AND PROFESSIONAL PRACTICE: TOWARDS EVIDENCE-BASED PRACTICE

Daniel L. Moody

Charles University Prague, Czech Republic
Monash University Melbourne, Australia

ABSTRACT

In most professional (applied) disciplines, research findings take a long time to filter into practice, if they ever do at all. The result of this is under-utilisation of research results and sub-optimal practices. There are a number of reasons for the lack of knowledge transfer. On the "demand side", people working in professional practice have little time available to keep up with the latest research in their field. In addition, the volume of research published each year means that the average practitioner would not have time to read all the research articles in their area of interest even if they devoted all their time to it. From the "supply side", academic research is primarily focused on the production rather than distribution of knowledge. While they have highly developed mechanisms for transferring knowledge among themselves, there is little investment in the distribution of research results beyond research communities. The World Wide Web provides a potential solution to this problem, as it provides a global information infrastructure for connecting those who produce knowledge (researchers) and those who need to apply this knowledge (practitioners). This paper describes two projects which use the World Wide Web to make research results directly available to support decision making in the workplace. The first is a successful knowledge management project in a health department which provides medical staff with on-line access to the latest medical research at the point of care. The second is a project currently in progress to implement a similar system to support decision making in IS practice. Finally, we draw some general lessons about how to improve transfers of knowledge from research and practice, which could be applied in any discipline.

KEYWORDS

knowledge management, evidence-based medicine (EBM), World Wide Web (WWW), IS research, IS practice, education, decision support system (DSS), web-based development

1. INTRODUCTION

The Problem Addressed

There is an enormous amount of new knowledge generated every year as a result of academic research. To make a practical difference, this knowledge needs to be disseminated and used in practice. Knowledge has no real value on its own¾it only becomes valuable when people use it make decisions and take action (Sveiby, 1997). However in most disciplines, research findings take a long time to filter into practice, if they ever do at all. There are a number of barriers to the flow of knowledge between research and practice, which originate from both sides of the divide:

- Practitioner's viewpoint (demand side): The pressures of professional practice leaves little time for practitioners to read journals or attend conferences: time is money, and most organisations do not reward their employees for keeping up with research in their field. Also, the volume of research published every year means that practitioners could not possibly keep up with all the latest research developments in their field¾if they did, they would have little time to do anything else. ·

- Researcher's viewpoint (supply side): Academic research is primarily focused on production rather than distribution of knowledge (Gibbons et al, 1994). Research communities have developed highly efficient mechanisms for transfer of knowledge among themselves (via the processes of publication and citation), but there is little investment in the dissemination of research results into practice. As a result, potentially valuable research ideas are circulated within research communities without ever finding their way into practice. Part of the reason for this is that academic institutions reward researchers for publishing their ideas in scholarly journals and conferences, not for having them applied in practice.

The result of these barriers is under-utilisation of research results and sub-optimal practices. This is undesirable from the point of view of practitioners and researchers:

- Researchers do not get their ideas tested in practice, which is a limiting factor in the development and evolution of these ideas (Wynekoop and Russo, 1997) ·
- In the absence of relevant knowledge about effectiveness of practices, practitioners persist in using practices that are obsolete or proven not to work. As a result, professional practice has limited ability to learn from its mistakes.

Manufacturing Model of Knowledge Production

In manufacturing, it is important to pay equal attention to production and distribution. To get maximum value from investments in production of goods, it is necessary to have parallel investments in distribution, to ensure that goods get sold and produce revenue. Similarly, to get maximum value from investments in research, it is necessary to have parallel investments in dissemination of research results, in order to improve practices and achieve social outcomes. The issue of how to transfer research results into practice is rarely addressed by researchers, and requires much more than publication in scholarly journals and conferences, which is normally seen as the endpoint of a research project. However, very little is known about how ideas are diffused in practice. The mechanisms are not as well understood as in the academic world, and rely more on informal channels, such as word of mouth (Gibbons et al, 1994).

Knowledge Management

Fundamental to understanding the issues involved in transferring research findings to practice is the concept of knowledge management. Knowledge management has only recently emerged as a field of practice in its own right (Sveiby, 1997; Davenport and Prusak, 1998). Knowledge is a high value form of information that can be used to make decisions and take action (Davenport et al, 1998). A key difference between knowledge and information or data is that it is intellectually intensive rather than IT-intensive. Knowledge is the result of human interpretation and analysis rather than data processing. Knowledge can be classified into two broad categories:

- Explicit knowledge: this is knowledge which has been articulated or written down. This is also referred to as "knowledge that", articulated knowledge or theoretical knowledge (Ryle, 1949; Polanyi, 1967; Rescher, 1979; Cohen and Squire, 1980; Cohen, 1984; Sveiby, 1997).
- Tacit knowledge: this is knowledge which is stored in people's heads. This is also referred to as "knowledge how", embodied knowledge or practical knowledge (Ryle, 1949; Polanyi, 1967; Rescher, 1979; Cohen and Squire, 1980; Cohen, 1984; Sveiby, 1997).

Explicit knowledge is the more familiar form of knowledge, and is found in textbooks, manuals and research articles. However tacit knowledge is more valuable because knowledge must be made tacit to make decisions or to take action. For example, you must read and understand a book (internalise the knowledge or make it tacit) in order to be able to apply the knowledge it contains. However because tacit knowledge is intangible, it is much more difficult to manage. Tacit knowledge can converted into explicit knowledge through the process of externalisation (or codification), and this has been the focus of many knowledge management initiatives in practice (Hansen et al, 1999). However, this is only ever partially successful, as we always know more than we can say¾also, some knowledge is not accessible to our consciousness (Polanyi, 1967; Sveiby, 1997). Formal education focuses on imparting explicit knowledge, while experience in practice supplements this with tacit knowledge through on-the-job learning and skills transfer (Sveiby, 1997). So far, knowledge management has been primarily concerned with managing knowledge flows within single organisations. However the same principles can be applied to the issue of improving knowledge transfers between research and practice within a discipline.

There are two key flows of knowledge which need to occur between research and practice in an applied discipline (Figure 1):

- Practice Research: research activity should be informed by the needs of practice and society. This ensures that research addresses issues that are of social and practical significance (relevance).
- Research Practice: research results should be disseminated and applied in practice. This ensures that research leads to

improvements in practices and benefits to society (impact). In any applied discipline, there is an expectation that research will ultimately result in some useful social outcome (Phillips, 1998).

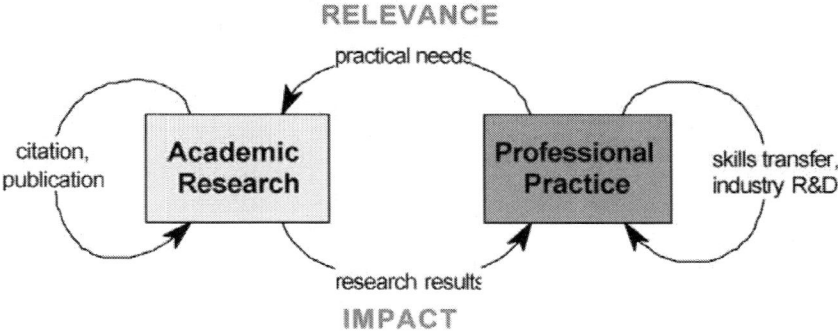

Figure 1. Knowledge Flows between Research and Practice

There are also flows of knowledge which occur within each community, although these are not the concern of this paper:

- The processes of citation and publication facilitate knowledge transfers among researchers.
- The process of skills transfer, in which tacit knowledge is transferred via on-the-job learning.
- Industry-based research and development, or "Mode 2" research, in which knowledge is generated and disseminated in the context of practice (Gibbons et al, 1994). This is characteristic of knowledge produced by pharmaceutical companies, consulting companies and IT vendors.

Improving knowledge transfers between research and practice will result in more effective production and application of research knowledge. In this paper, we focus on the flow of knowledge from research to practice, but this should also help to improve flows of knowledge in the reverse direction, via natural feedback processes.

Current Knowledge Distribution Model

Until now, there has been an implicit assumption that the best way to transmit research knowledge into practice is to first load it into human minds, via the long and expensive education of professionals (see Figure 2). However there are enormous "voltage drops" along this transmission line (Weed, 1997):

- Only a portion of the available knowledge can be loaded, given the limitations of human information processing and the limited time available for their education.
- Only a portion of the knowledge loaded is retained, and in fact most is quickly forgotten after examinations.
- Only a portion of the knowledge loaded is ever used. It is difficult to predict what knowledge might be useful in the future, so educators try to load as much as possible to prepare for all contingencies. This results in information overload and over-taxing of the retention powers of the human mind.
- Much of the knowledge that is retained quickly becomes obsolete, and there is no assurance that it will be replaced by relevant new knowledge. Once practitioners have completed their formal education, they tend to rely primarily on tacit knowledge acquired through experience in practice¾this means they are operating

from a knowledge base which is incomplete, out of date and biased.

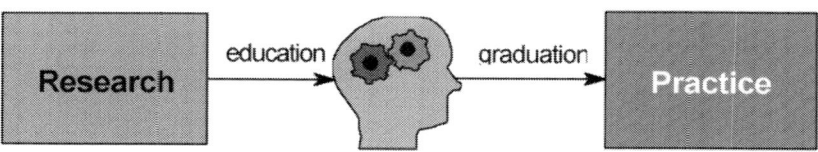

Figure 2. Current Knowledge Distribution Model

In the "Information Age", this seems a rather archaic and inefficient way of transferring knowledge. Just as science developed the microscope to magnify the power of the human eye, there is a similar need to use information technology to expand the human mind's limited capacity to store and recall large amounts of information (Weed, 1997). Practitioners should not need to load enormous amounts of information into their heads and then try to retrieve it years later (when it is likely to be out of date), but should have the latest research knowledge at their fingertips in the workplace. This will not substitute for the education process, but provides a way of continuously and selectively updating their knowledge. The education process only provides a "snapshot" of knowledge in a field at a point in time. Whenever a practitioner encounters a new situation, they need to be able to update their knowledge on an "as needed" basis¾this is the only workable approach given the rate of change in most disciplines.

Objectives of this Paper

Improving the transfer of knowledge from research to practice requires a channel which connects knowledge producers (researchers) with the intended consumers of this knowledge (practitioners). The World Wide Web provides a global information infrastructure for connecting researchers and practitioners. This paper describes two projects which use the World Wide Web to make research knowledge directly available to practitioners in the workplace. The first is a successful knowledge management project in a health department which provides medical staff with on-line access to the latest medical research at the point of care, in order to improve the quality of clinical decision making. The second is a project currently in progress to implement a similar system to support IS decision making. Finally, we draw some general lessons from these projects about transferring knowledge from research and practice, which could be usefully applied in any discipline.

Case Study 1: Evidence-Based Medicine

Medical Decision Making

To make appropriate decisions about patient care, medical practitioners need to take into account all relevant medical knowledge and integrate it with detailed data about the patient's condition. Access to the latest medical research can mean the difference between life and death, an accurate or erroneous diagnosis, early intervention or a prolonged and costly stay in hospital (Ayres and Clinton, 1997).

However, staying abreast of all the latest developments in medical research is a daunting task¾it is estimated that the amount of medical knowledge doubles every five years (Weed, 1997). Faced with information overload, doctors fall back on global judgements based on experience rather than thorough analysis of the relevant medical literature.

For decades, the medical profession has documented high levels of error, far higher than most other industries would tolerate. An adverse event is defined as "an unintended injury or complication which results in disability, death or extended hospital stay, and is caused by health care management rather than the patient's disease" (Wilson et al, 1995). In simple terms, this corresponds to an "error" in medical practice. A study of the Australian health system found that 16.6% of hospital admissions resulted in adverse events (Wilson et al, 1995). It is estimated that 180,000 people die each year in the United States as a result of adverse events-the equivalent of three jumbo jet crashes every day (Leape, 1994). When the causes of adverse events are investigated, it is found that most (over 80%) are the result of decision making errors (Wilson et al, 1999).

One of the reasons for the high levels of error in medical practice is its reliance on the unaided human mind (Weed, 1997). Psychological research shows that the human mind is strictly limited in how much information it can memorise and recall (Miller, 1956; Baddeley, 1994; 1999; Eysenck and Keane, 2000). In everyday medical practice, doctors rely heavily on tacit knowledge accumulated as a result of their medical training and subsequent clinical experience. As a result, decisions about the most appropriate treatment for patients are often made based on knowledge that is out-of-date, incomplete and biased. As Weed (1997) says:

"We should never have placed so much power in the hands of those who memorise knowledge, regurgitate it in examinations, forget much of it, specialise in a small piece of it, and never fully integrate what they do with the details of patient's needs."

Evidence-based Medicine

Medical research findings take a long time to filter into clinical practice (Phillips, 1998). Empirical studies have shown that on average, there is an 8 – 13 year time lag (depending on the specialty) between a treatment being proven to work and its adoption in everyday medical practice. It has also been found that 70% of medical treatments currently in use do not have any reliable evidence to support that they are any more effective than doing nothing (Sackett et al, 1997). The widespread use of disproven or unproven treatments means that the health system is burdened with unnecessary costs and patients are exposed to unnecessary pain, expense, risk and potential adverse side-effects.

Medicine has well-developed mechanisms for transferring tacit knowledge, via the internship process (on the job learning and skills transfer). However for transfer of explicit knowledge, it relies mainly on the formal education process, which as explained previously, is subject to severe "voltage drops". A major barrier to the implementation of research findings is the volume and geometric growth of the medical literature. It is not humanly possible to keep up with all the advances in all areas of medical research¾the average medical practitioner would have to read 20 articles per day just to keep up with developments in their specialty area (Jordens et al, 1998). It is also difficult for clinicians not trained in research methodology to make sense of the often conflicting research findings in a particular area.

Recognition of such problems led to the concept of evidence-based medicine (EBM). This is an approach that synthesises research findings on the effectiveness of medical treatments to support clinical decision making (Sackett et al, 1997). The purpose of EBM is to ensure that decisions about patient care are based on the latest scientific evidence, where "evidence" is defined as the results of randomised clinical trials (RCT). The aim is to reduce the time lag between the development of new treatments and their use in everyday medical practice, and to avoid the use of unproven or disproven treatments. The goals of EBM can be summarised as follows:

- To use treatments that have been proven to work
- To eliminate the use of treatments that have been proven not to work (disproven treatments)
- To conduct research into treatments whose effects are unknown (unproven treatments)

One of the major methodological tools in EBM is the systematic review, which is a form of meta-analysis of research results (Cochrane, 1972; Glass et al, 1981). Systematic reviews begin with an exhaustive search for published and unpublished research studies addressing a particular medical issue (e.g. treatment of asthma). The next step is to critically evaluate the studies to identify which are of sufficient quality to contribute to decision making. The final step is to pool the results of the studies to arrive at a quantitative estimate of the effectiveness of the treatment(s). Reviews must be regularly updated to take account of new research developments. This reduces the problems of information overload and interpretation of findings faced by medical practitioners, and puts information in a convenient form for decision making in practice.

However synthesising the research evidence is only the starting point for using research to improve practice. Equally important is the dissemination and use of this information. To make a practical difference, systematic reviews must be readily available to medical practitioners, and must be actively used in everyday clinical practice. EBM represents an application of knowledge management in the medical field, although it pre-dates the mainstream knowledge management literature by more than two decades. It focuses on synthesising explicit knowledge in the form of research findings, and disseminating and applying this knowledge in medical practice. It also focuses on generating new knowledge on the effectiveness of treatments.

Organisational Context

This case study took place in one of Australia's state health departments. The department is one of the largest organisations in Australia, with an annual budget of over AUD$8 billion and over one hundred thousand staff. Historically, clinical information needs had not been well supported by investments in information technology. The majority of existing information systems supported administrative processes (e.g. financial systems, payroll systems, patient administration systems), with very few systems directly supporting patient care. To address this issue, a committee was formed in September 1995 to specify requirements for medical information at the point of care. It consisted of fifty clinicians from all health disciplines including hospitals, general practice, community health and universities. A proposal was developed for a project called the Clinical Information Access Project (CIAP) to address this need, which was endorsed by senior management in December 1996. The stated objective of the project was:

"To provide clinicians with access to on-line medical information to support clinical practice, education and research at the point of care".

A pre-implementation survey of 2757 clinicians and medical librarians was carried out to identify the most important knowledge sources to support clinical practice. The system went live on July 4, 1997, taking just over six months from its initial inception to implementation. It has now been in operation for over five years, and is still evolving.

Knowledge Content

The knowledge content of the system was based on the requirements identified in the pre-implementation survey. It consists of five major components, which correspond to different classifications of knowledge on the tacit/explicit dimension and the internal/external dimension (whether the knowledge was produced inside or outside the organisation), as shown in Table 1. Unlike most knowledge management systems, which primarily focus on internal knowledge (Sveiby, 1997; Davenport et al, 1998; Davenport and Prusak, 1998; Hansen et al, 1999), this system is mainly focused on providing access to external knowledge, and in particular, the results of medical research.

Table 1. Knowledge Components

	EXTERNAL	INTERNAL
EXPLICIT	Systematic (EBM) Reviews On-line Literature Searching Tools Pharmaceutical Databases	Clinical Policies and Protocols
TACIT		Listservers

Systematic Reviews

A range of databases are provided that provide on-line access to systematic (EBM) reviews:

- Cochrane Library: this is an electronic database of systematic reviews produced by the Cochrane Collaboration, an international not-for-profit organisation specialising in principles of EBM. The Cochrane Library is recognised as the leading source of EBM reviews.
- APC EBM Reviews: systematic reviews produced by the American College of Physicians.
- Evidence-Based Medicine: systematic reviews produced by the British Medical Association.
- Database of Abstracts of Reviews of Effectiveness: systematic reviews produced by the British National Health Services' Centre for Reviews and Dissemination (NHS CRD).

These databases provide high value knowledge in the form of synthesised research findings. Each EBM review incorporates thousands of hours of critical review and synthesis of research articles by leading medical researchers and practitioners.

On-line Literature Searching Tools

Access to primary research sources (medical research journals) is also provided:

- MEDLINE: this is recognised as the leading bibliographic source of medical research, and provides on-line searching with links to full text medical journals.
- CINAHL: provides on-line searching with links to full text nursing and allied health journals.
- PsychINFO: provides on-line searching with links to full text psychiatry and psychology journals
- Healthstar: provides on-line searching with links to full text health administration journals
- Full text journals and textbooks: a range of full text journals and medical textbooks are available on-line

These databases provide access to the latest medical research findings. Compared to EBM reviews, they have the disadvantage that clinicians must synthesise and interpret the research findings themselves, but have the advantage of broader coverage (EBM reviews are only available for a limited range of topics) and currency (EBM reviews are only updated on an annual basis).

Pharmaceutical Databases

The system provides a range of on-line sources of drug information:

- MIMS: a comprehensive pharmaceutical database, which includes details of all known drugs, side-effects, interactions and recommended dosages.
- Antibiotic Guidelines: provides decision support in prescribing antibiotics (which are generally poorly prescribed in practice).
- Micromedex: poisons and toxins information.

Up-to-date drug information (e.g. recommended dosage, administration instructions, allergies, side-effects and interactions) is essential for prevention of adverse events. A recent study of adverse events in general practice found that over 50% were the result of pharmacological errors (Kidd and Veale, 1998). With thousands of drugs currently on the market, and hundreds of new ones released every year, it is impossible for medical practitioners to keep information about all of these in their heads (Milne, 2002).

Clinical Policies and Protocols

Clinical policies and protocols define standard procedures for handling particular types of cases (e.g. cardiac arrest, road trauma). These play a critical role in medical practice¾incorrect protocols or failure to follow protocols represent the most common cause of adverse events (Wilson et al, 1995; Wilson et al, 1999). They represent an important source of explicit knowledge in the organisation. Each hospital defines its own policies and protocols, and there may be wide variations between them based on equipment available and when they were last updated. The system allows clinicians to post their clinical policies and protocols on a voluntary basis for peer review. The purpose of this is to encourage collaboration and sharing of knowledge, and to move towards standardisation and development of "best practices".

Listservers

Listservers are provided for on-line discussion of problems and issues in particular areas of specialty, for example, asthma, stroke and medical ethics. This facilitates the exchange of ideas and experiences (tacit knowledge) among researchers and clinicians across the organisation. This is especially useful for clinicians in rural and remote areas, who have less opportunity for face-to-face exchange of knowledge. Strictly speaking, it is not possible to transmit tacit knowledge via a listserver¾the knowledge must be externalised or codified (transformed into explicit knowledge) before it can be communicated. However listservers provide a more informal and personal channel for knowledge transfer that enables communication of knowledge that would normally remain tacit.

Technology Architecture

The system operates using a single web server located at the organisation's head office, and is available 24 hours a day, 7 days a week. The World Wide Web was chosen instead of intranet technology in order to maximise the reach of the system. Because of the geographic spread of the organisation and the lack of a communications network linking all health care facilities, an intranet would have excluded a large proportion of clinicians. An added advantage of using the World Wide Web is that it allows clinicians to access the system from home, which represents around half the total usage of the system.

Scenario-How the System is used

To illustrate how the system is used in everyday medical practice, consider the case of a patient who presents with funnel web spider bite. The funnel web spider (Atrax Robustus) is unquestionably the most dangerous spider in Australia. It is a large (6-7 cm), black, aggressive spider, with fangs large and powerful enough to penetrate a fingernail. During a bite, the spider firmly grips its victim and bites repeatedly, and the venom is highly toxic. There is estimated to be 30-40 cases of funnel web spider bite occur each year in Australia. Given the rarity of such cases, it is unlikely that a clinician would be able to rely on previous experience or their medical training to know how to treat such a case. Death occurs between 15 minutes and 3 days following the bite, so prompt action is essential

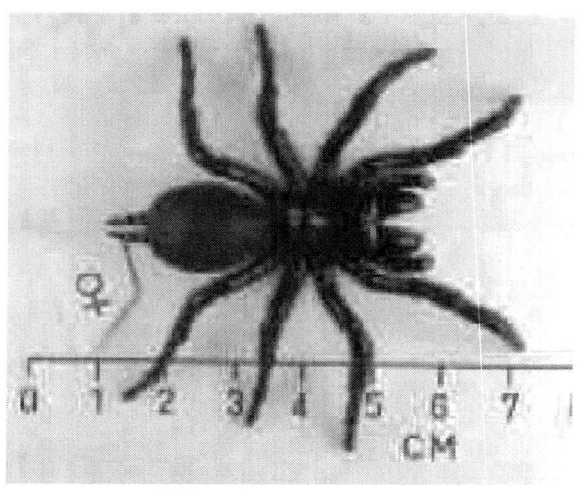

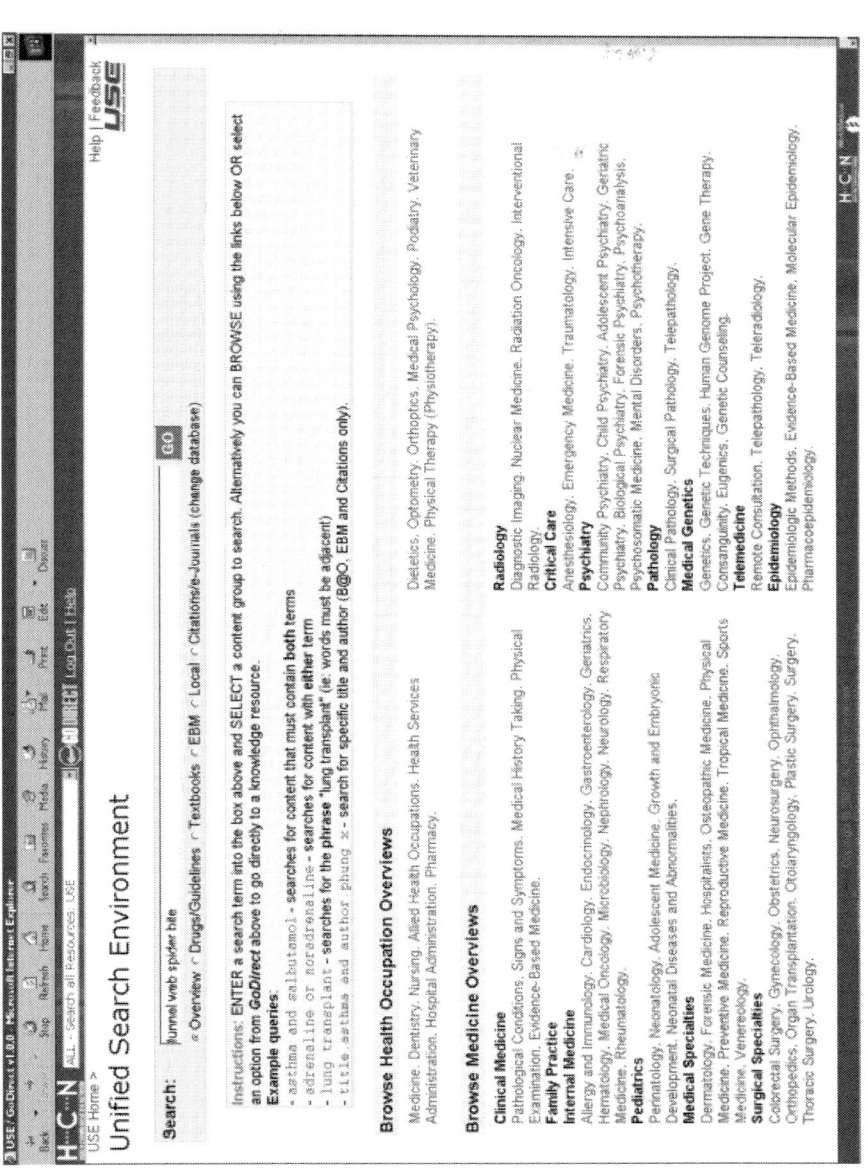

Figure 3. Unified Search Environment

From the opening screen, the attending medical officer conducts a search using the phrase "funnel web spider bite" (Figure 3). This initiates a search across all knowledge sources in the system.

The top-ranked result of the search is an entry from the Drug and Therapeutic guidelines, which describes the symptoms of funnel web spider bite (for confirming the diagnosis), the antivenom required, recommended dosage, administration instructions, interactions and possible side effects (Figure 4).

Benefits of the System

Both anecdotal evidence and perceptions of clinicians support the conclusion that the system has been successful in improving the quality of clinical decision making. A survey of clinicians using the system revealed that over 90% felt that it had improved patient care. There have also been a number of reported cases where access to the system has saved lives:

- A registrar in a rural hospital was able to save a patient in a critical condition suffering from the Lyssavirus. The Lyssavirus, which is acquired from contact with bats, causes encephalitis in humans and can be fatal if not treated quickly. However because it is a relatively new disease (the first case was discovered in Australia in May 1996), it does not yet appear in medical textbooks. Using MEDLINE, the clinician was able to look up and apply the appropriate treatment.

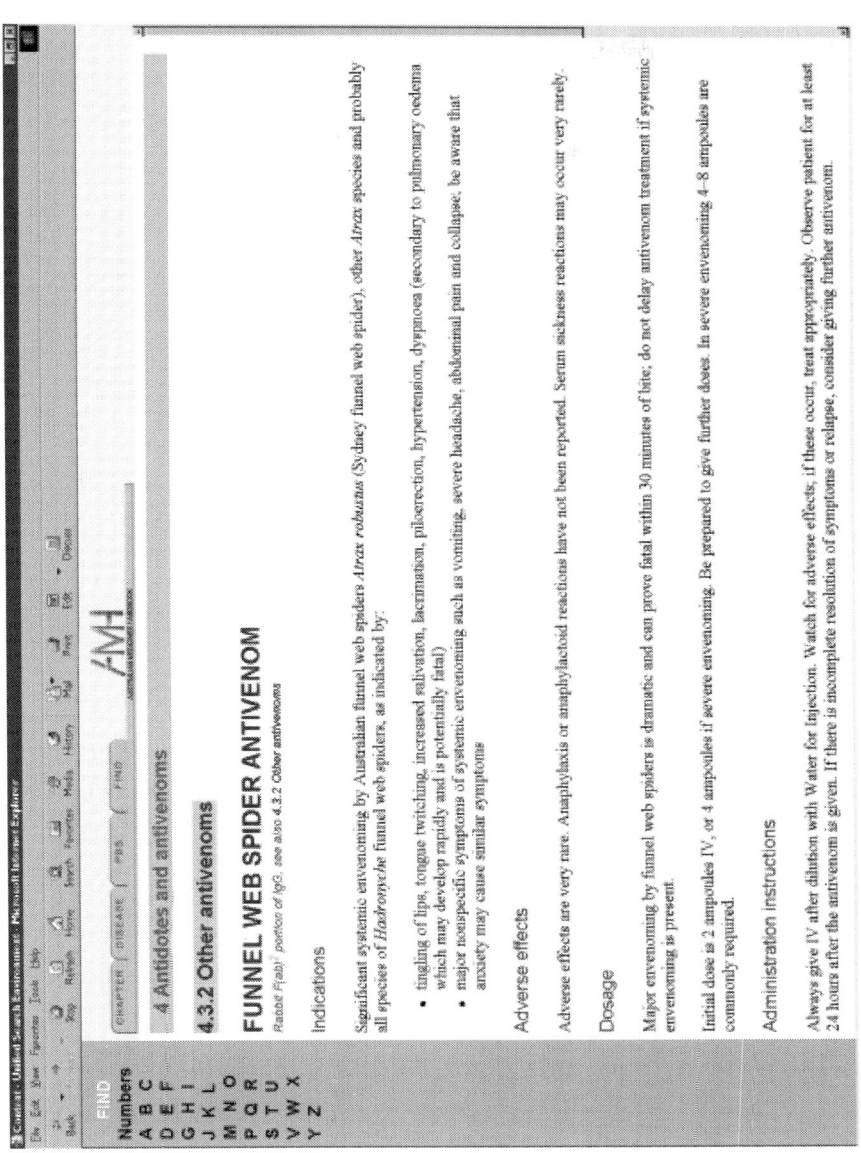

Figure 4. Search Results

- Another rural clinician was able to save a patient who was admitted in a critical condition in the middle of the night, suffering from meningitis. The patient did not respond to the normal treatment for this condition, indicated by the standard clinical protocol. A MEDLINE search revealed a new drug treatment that had immediate results.

- Finally, a clinician was able to save a child who had been bitten by a funnel web spider by looking up the anti-venom in Micromedex. Funnel web spider bite can be fatal, especially in children, and for the antivenom to be effective, it must be applied within 30 minutes. Having on-line access to this information almost certainly meant the difference between life and death.

Towards Evidence-Based Medical Practice

This system provides a mechanism for moving away from experience based medical practice (based on tacit knowledge) and towards evidence-based medical practice (based on explicit knowledge in the form of medical research). It provides a way of putting the theory of evidence-based medicine into practice, and using information technology to supplement the limited capability of the human mind to store and recall information. The system provides a direct channel for transfer of research results to clinical practice, and also directly connects researchers and practitioners via listservers. While the concept of EBM has been around for over three decades, the use of IT to provide on-line access to medical evidence in the workplace operationalises the concept in clinical practice.

Prior to this system, clinicians only had access to medical research via libraries attached to the larger hospitals. However these were generally not open after hours or on weekends and many smaller hospitals did not have libraries at all. Also, because of the pressures of clinical practice, medical practitioners rarely had the time to visit the library during working hours. Using this system, clinicians have on-line access to the latest medical research 24 hours a day, 7 days a week, both in the workplace and at home.

One of the explanations for the success of this system and its acceptance by clinicians is that unlike traditional decision support systems (DSS), which have proven to be largely unsuccessful in medical practice, it does not tell the clinician what to do. Twenty years of DSS research has revealed a persistent problem, in that decision makers either ignore or modify the advice given by the DSS (Lim and O'Connor, 1995; Turban, 1995; Lawrence and Sim, 1999; Goodwin, 2000). A range of studies in different environments have shown decision makers tend to trust their own experience and judgement much more than the DSS, even when presented with overwhelming evidence that they would perform much better by simply following the system's advice. A possible explanation for this is that people, and especially highly educated and experienced professionals, do not like being told what to do by a computer. The system described in this study supports human decision making by providing information about available treatments and the evidence for and against them. However it leaves the final decision about what to do to the clinician¾it therefore empowers human judgement and allows them to incorporate their experience and judgement into the decision making process.

Removing the Knowledge Gap between Practitioners and Consumers

If use of the system was extended to consumers, it would potentially have an even greater impact. In particular, patients could use it to look up the latest evidence about their own medical conditions. If they found that there was no reliable evidence to support use of the treatment they were being given, or that there was a better, clinically proven treatment available, they would rightly question their doctor's decision. The medical profession currently enjoys almost "god like" status in society, as a result of the wide knowledge gulf between medical practitioners and consumers. Patients are often uninformed, passive objects in the medical care process, and rarely question their doctor's advice (Weed, 1997). Making the latest research evidence available to patients would help to reduce the knowledge gulf between doctors and patients. While such a move might be resisted by medical practitioners¾as knowledge is power and most professional groups are concerned with maintaining their power and status¾it would help to improve accountability and the quality of health care.

Case Study 2: Evidence-based IS Practice

Decision Making in IS Practice

There is an enormous amount of IS research published each year, but very little of this knowledge finds its way into practice (Bubenko, 1986; Keen, 1991; Galliers, 1994; Benbasat and Zmud, 1999; Davenport and Markus, 1999; Moody, 2000). IS practitioners rarely refer to research evidence to make decisions but rely instead on their own experience, talking to peers, articles in the popular press or advice from consultants and vendors ("expert opinion"). However all of these knowledge sources are highly subjective and anecdotal. This means that, like medical practitioners, but to a much greater extent, they are operating from an incomplete and biased knowledge base.

Like medical practice, IS practice also exhibits high levels of errors. For example, empirical studies show that only 16.2% of software projects are completed on-time and on-budget (Standish Group, 1995). This represents a success rate about the same as the error rate in medical practice (cf. 16.6% of hospital admissions result in adverse events). Part of the reason for the high error rates in IS practice is the failure to learn from mistakes and the persistent use of practices which do not work. Research plays a vital role in enabling a profession to learn from its mistakes and in evaluating the effectiveness of practices.

The major reasons why IS practitioners do not refer to published research are:

- Time pressures: The constant deadline pressure and long working hours in IS practice leaves little or no time to read journals or attend conferences.
- Information overload: The volume of IS research published means it is not humanly possible for IS practitioners (and challenging even for full-time researchers) to keep up with all the relevant research in their area of practice. A recent study identified over 233 journals in the IS field (Hardgrave and Walstrom, 1997). The result is a situation of massive information overload for IS practitioners.

- Accessibility: the average IS practitioner would have great difficulty understanding papers published in the leading IS journals, because of the excessive focus on rigour and use of complex statistics and mathematics. IS research papers are primarily written for other researchers (conference and journal reviewers), which acts as a barrier to utilisation of research findings in practice (Moody, 2000).

A number of industry research organisations have emerged to support IT decision making in practice. However while this research is more accessible and relevant to practitioners, it is of questionable validity.

Towards Evidence-based IS Practice

The concept of evidence-based medicine—of making decisions based on the latest research evidence rather than on experience and opinions—is something that has clear application for the IS field. The IS field exhibits far more serious problems than medicine in transferring knowledge from research to practice, and the principles of EBM would help to bring research and practice closer together. There are important similarities between IS and medicine, which suggest that the concept can be adapted from one field to another. Both are applied rather than pure disciplines, which focus on applying technology to solve practical problems (Moody, 2000). While medical practitioners apply medical interventions to improve the health of their patients, IS practitioners apply IT-based interventions to improve organisational effectiveness. The problem of transferring research results to practice is essentially the same in both disciplines, though in IS the problem is much worse.

By analogy to evidence-based medicine, the objectives of evidence-based IS practice would be to:

- Use practices that have been proven to work
- Eliminate practices that do not work (disproven practices)
- Conduct research into practices whose effectiveness is unknown (unproven practices)

The third objective would provide direction for IS research in evaluating the effectiveness of practice, which would help to improve its relevance (the feedback loop in Figure 1).

The system described in the first case study provides a successful solution to the problem of transferring research knowledge to practice. It operationalises the concept of EBM in the form of a web-based information system. In this case study, we adapt this solution to the problem of transferring knowledge from IS research to IS practice. This is an example of analogical reasoning, a problem solving approach in which a solution is found to a problem by looking at how a similar problem has been successfully solved in another (referent) domain (see Figure 5). (Gentner, 1983; Holyoak, 1985; Keane, 1985; 1988). In this case:

- The target domain is IS, while the referent domain is medicine;
- The problem to be solved is that of transferring IS research knowledge to IS practice, while the exemplar problem is the transfer of medical research knowledge to medical practice;
- The exemplar solution is the webbased system described in the first case study, while the target solution is a similar system to support evidence-based IS practice.

Analogical reasoning represents a particularly useful research approach for an immature discipline like IS, and a way of learning from other, more established disciplines (in this case, medicine).

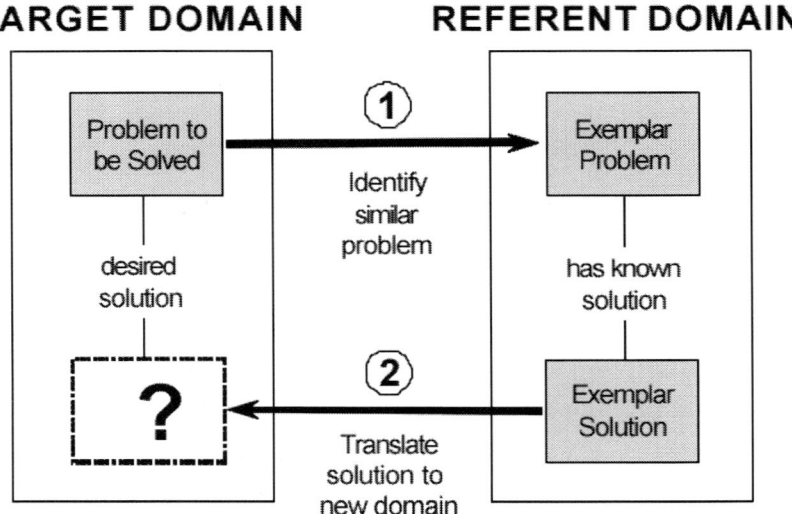

Figure 5. Analogical Reasoning as a Problem Solving Approach

Project Objectives

The objective of this project is to develop a web-based system to provide IS practitioners with access to the latest IS research, to support decision making in the workplace. This is an ambitious project, which attempts to build a comprehensive knowledge base to support IS practice. This is an application of principles of knowledge management on a large scale—the level of an entire discipline—and is a joint venture between a research institution (Monash University) and the association for IS professionals in Australia (the Australian Computer Society).

The design of the system will be closely based on the system described in the first case study. The system will incorporate the following components:

- On-line literature searching: this will provide sophisticated searching of IS journals with links to full text journals. This facilitates the transfer of explicit knowledge between research and practice. A number of existing on-line literature search engines could be used for this purpose. While such search engines are standard "tools of the trade" for IS researchers, few practitioners have access to them in the workplace.
- Systematic reviews: this will involve the formation of expert committees, consisting of both researchers and practitioners, to review research findings and synthesise results on particular topics. This is a value-added process which will put research results in a highly synthesised form to support decision making in practice. This is a process of synthesising explicit knowledge and provides a way of making research results more accessible to practitioners and reducing problems of information overload (Atkins and Louw, 2000; Moody, 2000).
- Posting of policies, standards, architectures, software evaluations etc. by practitioners for peer review: this would be done by subscribers on a voluntary basis to facilitate knowledge sharing, development of "best practices" and industry standardisation. This facilitates sharing of explicit knowledge among practitioners. Commercial interests will be a limiting factor in such knowledge

sharing, as many organisations will want to protect their intellectual property and competitive interests, but in many areas, there will be clear mutual benefits in sharing knowledge.

- Listservers to promote discussion on particular issues: this would facilitate transfer of tacit knowledge in the form of ideas and experiences between practitioners and researchers.

The intention is to make the system available free of charge to all members of the Australian Computer Society

Potential Benefits

Implementation of this system will help to:

- Improve the dissemination and application of IS research results in practice
- Reduce time lag between development of new research knowledge and its application in practice
- Help to reduce the use of practices that don't work and thereby facilitate learning from mistakes
- Connect researchers and practitioners together (via listservers)
- Provide a comprehensive and up-to-date knowledge base to support decision making in IS practice
- Improve the professionalism of IS practice, by moving towards evidence-based rather than opinion and experience based practice
- Reduce problems of information overload for IS practitioners and make it easier for them to stay abreast of the latest research (via systematic reviews)
- Improve sharing of knowledge among practitioners (via posting of policies, standards etc.) As in the previous study, this system

would also be beneficial to consumers of IS services. Many end users are relatively uninformed and rely heavily on advice from vendors and consultants which is often biased (and sometimes quite predatory). Consultancy firms and IT vendors are notorious for taking advantage of users' lack of knowledge to sell them products or services that they don't need or are not effective. Making the latest IS research available to consumers of IS products and services will help to ensure that they get better value for money and also improve accountability of the IS profession.

Potential Barriers

There are also a number of potential barriers to successful implementation of this system, which reflect some fundamental differences between medical and IS practice:

- Scepticism about IS research: IS practitioners tend to be rather sceptical of the value of academic research. This would be a significant barrier to adoption and use of the system. However if the system is found to be useful in practice, and a critical mass of practitioners start using it, normative influences are likely to overcome this scepticism (Green, 1998).

- Lack of evidence: in most areas of IS practice, there isn't a sufficient body of research to conduct effective systematic reviews. Because of the rate of technological change, it is difficult for IS research to keep pace with new technological developments and evaluate them in a timely manner. In addition, the conclusions from IS research studies tend to be much weaker than those in medicine. Because most IS interventions are applied at the organisational level, randomised clinical trials, which are the standard method for obtaining evidence in medicine, are not as widely applicable¾in fact, some researchers have argued that experimental methods are inappropriate in IS research. IS research relies much more heavily on qualitative research methods, which leads to less definite conclusions about effectiveness of practices.

- Relevance of IS research: A number of authors have criticised IS research for not addressing questions that are relevant to practice (Keen, 1991; Galliers, 1994; Benbasat and Zmud, 1999; Davenport and Markus, 1999). This is a "supply side" issue that is difficult to address in the short term. However as research results are more widely disseminated in practice, relevance of research should be improved through natural feedback mechanisms. In addition, listservers provide a direct channel for practitioners to influence research priorities.

- Need to incorporate non-academic research sources: Gibbons et al (1994) argue that universities no longer have a monopoly on knowledge production, and that especially in high tech fields, most new knowledge originates from practice than from academic research. This suggests that it is important to include industry based knowledge sources rather than restricting knowledge sources to academic research.

2. CONCLUSION

2.1 Summary of Findings

For research to make a practical difference, research results must be readily available to practitioners, and must be actively used and implemented in everyday practice (Jordens et al, 1998; Phillips, 1998). The World Wide Web provides an infrastructure for connecting the producers of research knowledge (researchers) with the intended consumers of this knowledge (practitioners). This paper has described a successful system which provides on-line access to the latest medical research to support clinical decision making at the point of care. It also describes a project currently in progress to develop a similar system to support IS practice. Neither of these projects would have been possible prior to the development of the World Wide Web, and illustrate how it has the potential to improve practices and also to allow consumers to become more informed. Providing this information to consumers will erode the power base of practitioners, which they are likely to resist, but ultimately it is good for society and will be a powerful impetus towards improving practices.

Towards a Paradigm for Evidence-based Practice

The systems movement was founded with the objective to "investigate the similarities of concepts, laws and models from various fields and to help in useful transfers from one field to another" (von Bertalanffy, 1968). Systems theory uses the process of generalisation as an approach to achieving such transfers. A problem in a particular domain can be generalised to a systems problem by removing all domain specific aspects (Klir, 1985). The solution to the systems problem can then be applied in a wide range of domains.

Evidence-based medicine (EBM) is an approach for improving the effectiveness of medical practice. More generally, it provides a conceptual framework for improving knowledge transfers from research to practice. The problems addressed by EBM are experienced to a greater or lesser extent in all applied disciplines, and similar principles could be applied in any field to improve knowledge transfers from research to practice. The principles of EBM can be generalised to the systems level to develop a general paradigm for evidence-based practice, which can then be applied in multiple domains (Figure 6).

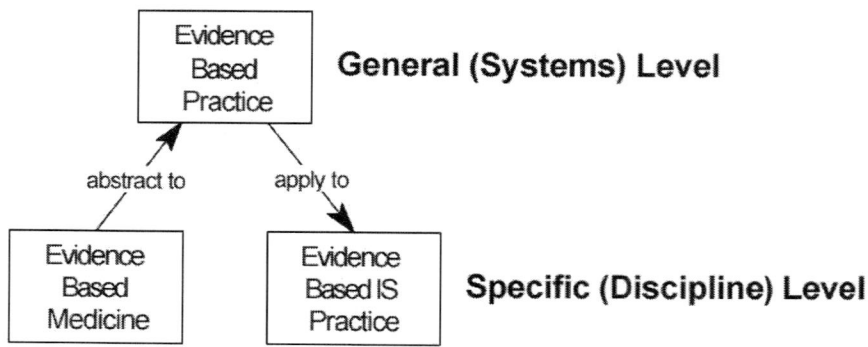

Figure 5. Evidence-based Practice

The two case studies described in this paper represent practical solutions to the problem of transferring knowledge from research to practice, which operationalise the concept of evidence-based practice in the form of information systems. The key elements of these systems are:

- Systematic reviews: synthesis of research literature to increase accessibility to practitioners and reduce time required to find and evaluate evidence.
- On-line literature searching tools: access to primary research sources with value-added search capabilities
- Posting of practitioner-developed "knowledge products" (e.g. policies, standards, procedures): sharing of explicit knowledge to support development of "best practices" and achieving industry standardisation.
- Listservers: informal channels for sharing knowledge between researchers and practitioners.

These systems demonstrate how information technology can be used to expand the human mind's limited capacity to store and recall large amounts of information. Providing practitioners with access to the latest research in the workplace provides a form of "just-in-time" learning, which allows them to selectively update their knowledge on an "as needed" basis. This will help to supplement the education process, which provides a "snapshot" of knowledge in a field at a point in time.

REFERENCES

1. Atkins, C. & Louw, G. (2000). Reclaiming knowledge: The case for evidence based information systems. Proceedings of the 8th European Conference on Information System (ECIS2000). Vienna, Austria.

2. Ayres, D.H.M. & Clinton, S. (1997). The user connection: Making the clinical information systems vision work in NSW health. Health Informatics Conference (HIC '97). Sydney, Australia.

3. Baddeley, A.D. (1994). The magical number seven: Still magic after all these years? Psychological Review, 101, 2.

4. Baddeley, A.D. (1999). Essentials of human memory. Hove, England: Psychology Press.

5. Benbasat, I. & Zmud, R.W. (1999). Empirical research in information systems: The practice of relevance. MIS Quarterly, 23, 1, March.

6. Bubenko, J.A. (1986). Information systems methodologies - A research view. Information Systems Design Methodologies: Improving the Practice, T.W. Olle, H.G. Sol and A.A. Verrijn-Stuart, (Eds.), North-Holland.

7. Cochrane, A.L. (1972). Effectiveness and efficiency: Random reflections on health services. London: Royal Society of Medicine Press.

8. Cohen, N.J. (1984). Preserved learning capacity in amnesia: Evidence for multiple memory systems. In L. Squire & N. Butters (Eds.), Neuropsychology of Memory. New York: Guilford Press.

9. Cohen, N.J. & Squire, L.R. (1980). Preserved learning and retention of pattern analysing skill in amnesia using perceptual learning. Cortex, 17, pp. 273-278.

10. Davenport, T.H., De Long, D.W. & Beers, M.C. (1998). Successful knowledge management projects. Sloan Management Review, 39, 2, pp. 43-52.

11. Davenport, T.H. & Markus, M.L. (1999). Rigour vs. relevance revisited: Response to Benbasat and Zmud. MIS Quarterly, 23, 1, March.

12. Davenport, T.H. & Prusak, L. (1998). Working knowledge: How organisations manage what they know. Boston, MA: Harvard Business School Press.

13. Eysenck, M.W. & Keane, M.T. (2000). Cognitive psychology: A student's handbook. Hove, England: Lawrence Erlbaum

14. Galliers, R.D. (1994). Relevance and rigour in information systems research: Some personal reflections on issues facing the information systems research community. Proceedings of the IFIP TC8 Conference on Business Process Reengineering: Information Systems and Challenges. Gold Coast, Australia.

15. Gentner, D. (1983). Structure-mapping: A theoretical framework for analogy. Cognitive Science, 7.

16. Gibbons, M., Limoges, C., Nowotny, H., Schwartzman, S., Scott, P. & Trow, M. (1994). The new production of knowledge: The dynamics of science and research in contemporary societies. Sage Publications.

17. Glass, G.V., McGaw, B. & Smith, M.L. (1981). Meta analysis in social research. London: Sage

18. Goodwin, P. (2000). Improving the voluntary integration of statistical forecasts and judgment. International Journal of Forecasting, 16, pp. 85-99.

19. Green, C. (1998). Normative influence on the acceptance of information technology. Small Group Research, 29, 1, February.

20. Hansen, M., Nohria, H. & Tierney, T. (1999). What's your strategy for managing knowledge? Harvard Business Review, March/April.

21. Hardgrave, B.C. & Walstrom, K.A. (1997). Forums for MIS scholars. Communications of the ACM, 40, 11, November.

22. Holyoak, K.J. (1985). The pragmatics of analogical transfer. The Psychology Of Learning And Motivation, 19.

23. Jordens, C.F.C., Hawe, P., Irwig, L.M., Henderson-Smart, D.J., Ryan, M., Donoghue, D.A., Gabb, R.G. & Fraser, I.S. (1998). Use of systematic review of randomised trials by Australian neonatologists and obstetricians. Medical Journal of Australia, March 16.

24. Keane, M. (1985). On drawing analogies when solving problems: A theory and test of solution generation in an analogical problem solving task. British Journal Of Psychology, 76.

25. Keane, M. (1988). Analogical problem solving. New York:Wiley.

26. Keen, P.G.W. (1991). Relevance and rigour in information systems research: Improving quality, confidence, cohesion and impact. In H.-E. Nissen, H.K. Klein & R. Hirschheim, (Eds.), Information Systems Research: Contemporary Approaches and Emergent Traditions. North Holland: Elsevier Science.

27. Kidd, M.R. & Veale, B.M. (1998). How safe is Australian general practice and how can it be made safer? Medical Journal of Australia, July 20.

28. Klir, G.J. (1985). Architecture of systems problem solving. New York: Plenum Press.

29. Lawrence, M.J. and SIM, W. (1999). Prototyping a financial DSS. Omega, 27, 4, pp. 445-450.

30. Leape, L. (1994). Error in medicine. Journal of the American Medical Association (JAMA), 272.

31. Lim, J. & O'Connor, M. (1995). Judgemental adjustment of initial forecasts: Its effectiveness and biases. Journal of Behavioural Decision Making, 8, pp. 149-168.

32. Miller, G.A. (1956). The magical number seven, plus or minus two: Some limits on our capacity for processing information. The Psychological Review, March.

33. Milne, E.B.G.W.A. (2002). Drugs: Synonyms & properties (2nd ed.). Burlington, VT: Ashgate.

34. Moody, D.L. (2000). Building links between IS research and professional practice: Improving the relevance and impact of IS research. In R.A. Weber & B. Glasson, (Eds.), International Conference on Information Systems (ICIS'00), Brisbane, Australia, December 11-13.

35. Phillips, P.A. (1998). Disseminating and applying the best evidence. Medical Journal of Australia, March.

36. Polanyi, M. (1967). The tacit dimension. London: Routledge & Kegan-Paul.

37. Rescher, N. (1979). Cognitive systematization. Oxford: Basil Blackwell.

38. Ryle, G. (1949). The concept of mind. Chicago: University of Chicago Press.

39. Sackett, D.L., Richardson, W.S., Rosenberg, W. & Haynes, R.B. (1997). Evidence based medicine: How to practice and teach EBM. New York: Churchill Livingstone.

40. Standish Group (1995). The CHAOS report into project failure. The Standish Group International Inc. Available on-line at http://www.standishgroup.com/visitor/chaos.htm

41. Sveiby, K.-E. (1997). The New organisational wealth: Managing and measuring knowledge-based assets. San Francisco: Berret-Koehler.

42. Turban, E. (1995). Decision support and expert systems: Management support systems, (4th ed.). Engelwood Cliffs, NJ: Prentice Hall.

43. Von Bertalanffy, L. (1968). General systems theory: Foundations, development, applications. Braziller.

44. Weed, L.L. (1997). New connections between medical knowledge and patient care. British Medical Journal, 315, July 26.

45. Wilson, R.M., Harrison, B.T., Gibberd, R.W. and Hamilton, J.D. (1999): "An analysis of the causes of adverse events from the quality in Australian health care study", The Medical Journal of Australia, May 3.

46. Wilson, R.M., Runciman, W.B., Gibberd, R.W., Harrison, B.T., Newby, L. & Hamilton, J.D. (1995). The quality in Australian health care study. The Medical Journal of Australia, November 6.

47. Wynekoop, J.L. & Russo, N.L. (1997). Studying systems development methodologies: An examination of research methods. Information Systems Journal, 7, 1, January.

4

CHAPTER

LINKING RESEARCH TO PRACTICE: THE RISE OF EVIDENCE-BASED HEALTH SCIENCES LIBRARIANSHIP*

Joanne Gard Marshall, PhD, AHIP, FMLA

School of Information and Library Science, University of North Carolina at Chapel Hill, 100 Manning Hall, Chapel Hill, NC 27599-0001

Purpose: The lecture explores the origins of evidence- based practice (EBP) in health sciences librarianship beginning with examples from the work of Janet Doe and past Doe lecturers. Additional sources of evidence are used to document the rise of research and EBP as integral components of our professional work.

Methods: Four sources of evidence are used to examine the rise of EBP:

(1) a publication by Doe and research-related content in past Doe lectures,

(2) research-related word usage in articles in the *Bulletin of the Medical Library Association* and *Journal of the Medical Library Association* between 1961 and 2010,

(3) Medical Library Association activities, and (4) EBP as an international movement.

Results: These sources of evidence confirm the rise of EBP in health sciences librarianship. International initiatives sparked the rise of evidence-based librarianship and continue to characterize the movement. This review shows the emergence of a unique form of EBP that, although inspired by evidence-based medicine (EBM), has developed its own view of evidence and its application in library and information practice.

Implications: Health sciences librarians have played a key role in initiating, nurturing, and spreading EBP in other branches of our profession. Our close association with EBM set the stage for developing our own EBP. While we relied on EBM as a model for our early efforts, we can observe the continuing evolution of our own unique approach to using, creating, and applying evidence from a varietyof sources to improve the quality of health information services

1. BACKGROUND

When I was first told that I had been selected as the Janet Doe Lecture for 2013, my reaction can best be described as a mixture of total panic and excitement. The panic was exacerbated when I began to read earlier Doe lectures, a practice that has become a ritual for the Doe lecturer-elect. I was able to revisit many of the Doe lectures that inspired my own career over the years and to read others that were less familiar to me. All contained useful insights as well as many visionary ideas. I was impressed by the "aerial view" that so many of the Doe lecturers brought to their topics. The ability to see the view from the top was clearly based on their years of involvement in the field and the kind of wisdom that can only come from having been there.

The excitement started to set in when I realized that One Health was a joint meeting of the 2013 Annual Meeting and Exhibition of the Medical Library Association (MLA '13), the 11th International Con- gress on Medical Librarianship (ICML), the 7th International Conference of Animal Health Informa- tion Specialists (ICAHIS), and the 6th International Clinical Librarian Conference (ICLC). As a British, Canadian, and, now, US citizen, I was reminded of how excited I had been when the world of medical librarianship opened to me in 1970, when I was hired as a librarian at the McMaster University Health Sciences Center. I had the good fortune to have as library director, Beatrix Robinow, who had been an MLA Cunningham Fellow from South Africa. She had subsequently moved to Canada and was the founding librarian at the new medical school at McMaster. Mrs. Robinow, as we always called her, was devoted to the Medical Library Association (MLA), serving on the Board of Directors from 1978 to 1981 and making sure that her fledging librarians were able to attend MLA meetings. I learned so much from Mrs. Robinow and MLA over the years, taking as many continuing education courses as I could and eventually teaching some myself. This personal history made the idea of presenting a Doe lecture at an international meeting especially exciting.

As I reflected on the aspects of the field that I had become passionate about, the idea of linking research to practice kept resurfacing: it seemed to sum up much of what I have been trying to do over the years. The advent of evidence-based library and information practice (EBLIP) and the leadership role that health sciences librarians were playing in its development seemed to be the perfect way to illustrate the practical value of linking research to practice. What a relief—at least I had a topic!

At that point, I thought I should learn more about the history of the Janet Doe Lecture and Janet Doe herself, so that I could get some ideas on how to take my topic and turn it into a respectable lecture. To my delight, I came across an article by Janet Doe in the *Bulletin of the Medical Library Association (BMLA)*, "The Survey and After" [1]. Although published in 1961, the article reported on a "survey" of the ailing Army Medical Library that was conducted in 1943 by a committee that included three medical librarians: Janet Doe, Mary Louise Marshall, and Thomas P. Fleming. In what researchers today would be more likely to call a detailed case study, the committee members conduct- ed a careful and thorough investigation of the state of the Army Medical Library. Their research methods included interviews, documentary evidence, and in- depth observation of the library site, its collections, and its services. Despite its reputation as "the greatest collection of medical literature in existence," stated Doe, the library and its collection were in dire straits after the Great Depression, housed in dilapidated quarters, with a drastically diminished budget and staff. She went on to describe the results of their study and its effect on the rebuilding of the library services and collections over the ensuing sixteen years, culmi- nating in the transfer of stewardship of the library to the new National Library of Medicine building in 1956. I was left to wonder if there had ever been a better example of how research can lead to profoundly consequential outcomes for our field.

2. THE RISE OF EVIDENCE-BASED MEDICINE

Next, I turned my attention to the rise of evidence- based medicine, since medicine was the first health discipline to adopt the evidence-based practice model. I experienced some of this early history firsthand when I was hired as a medical librarian at McMaster in 1970. The medical school was brand new, and among the faculty recruits was Dr. David Sackett, who came to head the Department of Clinical Epidemiology and Biostatistics. At McMaster, it was not going to be business as usual, and Dr. Sackett fit right in. He wanted to create a different kind of department that would link research to clinical practice. None of us in the library were quite sure how this would work, but Dr. Sackett's enthusiasm was contagious, and soon we were trying to find ways of supporting the department.

We developed a clinical librarian program to respond to information needs in the 450-bed hospital that was part of the academic health sciences center. This was where I caught my first glimpse of the impact of linking research to practice. One morning when I went to rounds, I saw the residents poring over a paper copy of an article I had put up on the bulletin board in the conference room, and I realized that they were going to change the care of a patient based on the results of the study reported in the article. I was hooked! Eventually, thanks to the advice and mentorship of Dr. Sackett and his colleagues, I was able to conduct a randomized controlled trial to evaluate the educational impact of our clinical librarian service [2].

Medical education at McMaster was problem based, which meant that students had to research a hypothetical patient case and apply what they had learned to solving the patient's clinical problems. Eventually, this process was applied to actual patient care. Lectures were replaced by tutorials and prob- lem-solving groups. The library and its resources were very much at the heart of the clinical problem solving activity. In retrospect, Dr. Sackett and his colleagues at McMaster medical school were sowing the seeds of what later became known as evidence- based medicine, although that was not the original terminology that was used.

In 1981, the Department of Epidemiology and Biostatistics began producing a series of articles in the *Canadian Medical Association Journal* on how to read clinical journals [3]. At McMaster, tutorials were organized on this topic with the title, "Critical Appraisal of the Literature." The librarians were involved along with faculty as mentors in the critical appraisal tutorials, and medical students were taught how to find evidence in the medical literature and apply it to patient care. A similar series of articles on critical appraisal, called "Users' Guides to the Medical Literature," began in the *Journal of the American Medical Association (JAMA)* in 1993 [4]. Dr. Sackett and his colleagues went on to publish a book on clinical epidemiology in 1985 that elaborated on his vision for linking research to practice [5].

In a 1996 article in the *British Medical Journal*, Sackett and his colleagues provided what had become the standard definition of evidence-based medicine: "The conscientious and judicious use of current best practice in making decisions about the care of individual patients" [6]. The evidence-based model consists of a combination of best research evidence from the research literature combined with clinical expertise and patient values and preferences. Integrated into this evidence- based model was a hierarchy of levels of evidence starting with expert opinion, followed by case report, case control studies, cohort studies, randomized con- trolled trials (RCTs), and systematic reviews. Each successive level in the hierarchy was considered to be a stronger form of evidence.

Evidence-based medicine has since spread world- wide, thanks to organizations such as the internation- al Cochrane Collaboration that bring together the best evidence from the medical literature. In the United States, the Agency for Healthcare Research and Quality (AHRQ) has funded the creation of systematic reviews and developed a repository for them. Over the years, evidence-based practice as a concept has spread to the other health professions as well as to fields outside of the health sciences, such as social work, education, and management.

Of course, there is more to the early development of evidence-based medicine than I have been able to describe in this lecture; however, I hope that this glimpse into its early development at McMaster and the important role played by librarians provides some useful insights.

3. SOURCES OF EVIDENCE FOR THE RISE OF EVIDENCE-BASED HEALTH

3.1 Sciences Librarianship

The Janet Doe Lectures

While it has taken some time for the concept of evidence-based practice to take hold in librarianship, we can see the seeds for this development not only in the work of Janet Doe, but also in the content of earlier Doe lectures. In 1977, Erich Meyerhoff, AHIP, FMLA, noted a shift from historical to scientific inquiry in the profession and cited the pool of talent represented by hospital librarians and the rise of the women's movement as instrumental factors in the change [7]. In her 1985 lecture, Lucretia W. McClure, AHIP, FMLA, discussed "The Promise of Fruit...and Light," nothing the possibility of changing our designation as librarians from "Keeper of the Printed Book" to "Keeper of

Knowledge" [8]. McClure also cited Estelle Brodman's article on citation patterns in physiology journals published in 1944, in which the author used her own research experience to create a critical appraisal of citation analysis as a research method [9]. Another early evidence-based practitioner!

In 1986, Doe lecturer Virginia H. Holtz, AHIP, FMLA, noted an enduring concern in the profession with "measures of excellence" including standards, library statistics, and other forms of data used to monitor and improve library services [10]. In 1987, Erika Love, FMLA, chair of the original MLA Research Committee, made direct reference to the importance of research to the profession in her address, "The Science of Librarianship: Investing in the Future" [11]. In 1989, Rachael K. Anderson, AHIP, FMLA, went on to cite "research competence' as one of the six key attributes required by librarians in their evolving roles [12]. She described roles both in assisting library users to do their research and conducting library research that will inform the development of library services. In her 1994 lecture, "The Idea of the Library," Nina W. Matheson, AHIP, FMLA, declared that "organizations will flourish who are able to apply knowledge to create knowledge and to organize it to produce knowledge" [13]. In 1998, Wayne J. Peay, FMLA, revisited the need for better data to inform library practice in his paper, "Strate- gies and Measures for the Next Century" [14].

The discussion of evidence-based practice has continued in some of the more recent Doe presenta- tions as well. In 1999, Sherrilynne Fuller, FMLA, was very direct in her reference to research when she spoke of "Enabling, Empowering, Inspiring: Research and

Mentorship throughout the Years" [15]. In 2005, Fred W. Roper, AHIP, FMLA, gave a history of the MLA continuing education program that reminded us of the importance of this program in our professional development, including the development of our research competencies [16]. In 2011, T. Scott Plutchak, AHIP, FMLA, used his experience as editor of the *Journal of the Medical Library Association (JMLA)* and in scholarly publishing to speculate on the "Dawning of the Great Age of Librarians" [17]. Plutchak noted that there is much work to be done by librarians in making the products of research accessible and usable to our library users and to ourselves. Despite the increas- ing abundance of research and scholarship, linking research to practice continues to be a challenge. Mark E. Funk, AHIP, FMLA, provided an illuminating look into our own changing world of practice through his textual analysis of articles published in the *BMLA* and *JMLA* from 1961 to 2010 [18]. I will leave further discussion of Funk's findings to the next section of this lecture.

Our own words

Fortunately for me, 2012 Doe Lecturer Funk created the perfect opportunity to study the rise of evidence- based practice in health sciences librarianship [18]. Funk's amazing feat of downloading articles pub- lished in the *BMLA* and *JMLA* between 1961 and 2010 resulted in an electronic corpus of 84,648 unique words that allowed him to track the frequency of word usage over time. His analysis revealed changes in 4 major areas: the environments in which we exist both inside and outside the library; our approaches to library and information management; the growing importance of technology, including digitization and the Internet; and an increasing interest and involve- ment in research. Using this dataset, I was able to delve more deeply into the use of research and evidence-based practice word usage. With the help of my research assistants, we identified 6 broad research terms in the dataset and the term variants related to them. The terms were as follows: "research" with 88 term variants, "survey" with 32 term variants, "evaluation" with 18 term variants, "methods" with 6 term variants, "evidence" with 19 term variants, and "measurement" with 13 term variants. A variant was defined as a different form of the word such as research, researcher, research-based, and so on. Between 1961 and 2010, there were 39,211 uses of all the research-related terms or their variants.

While word usage increased for all of the term groupings, the terms research and evidence and their variants show the most rapid rise, especially since 1990. Figure 1 provides a visual picture of the rise in each decade as well as the relative frequency of term use. The popularity of the survey as a form of data collection is evident, as is the focus on evaluation as a type of research. A similar decade-by-decade analysis of word use as a percent of all words in the database showed a similar trend.

Funk made effective use of sparklines, which are helpful for visualizing patterns in time series data such as the *BMLA/JMLA* corpus. The reader is referred to Funk's article for a discussion of the origin and use of this method for displaying data [18]. Using Funk's approach, the following sparklines show the actual rise and fall of word usage each year as well as a superimposed trend line, based on the data over time. Sparklines were developed for each of the six broad research terms. In Figure 2, the continuous rise in use of the term research and its variants is evident. In Figure 3, there is a steady but still rising curve for the term evaluation and its variants. The use of methods terms shown in Figure 4 also shows a consistent rise. In Figure 5, the use of the term evidence and its variants was low and flat from the 1960s through the mid-1990s, but then it started to rise quickly. The two remaining sparklines for the terms survey and measure(ment) (Figures 6 and 7) show the continuing popularity of the survey method for collecting data but more modest gains for terms related to measure(ment).

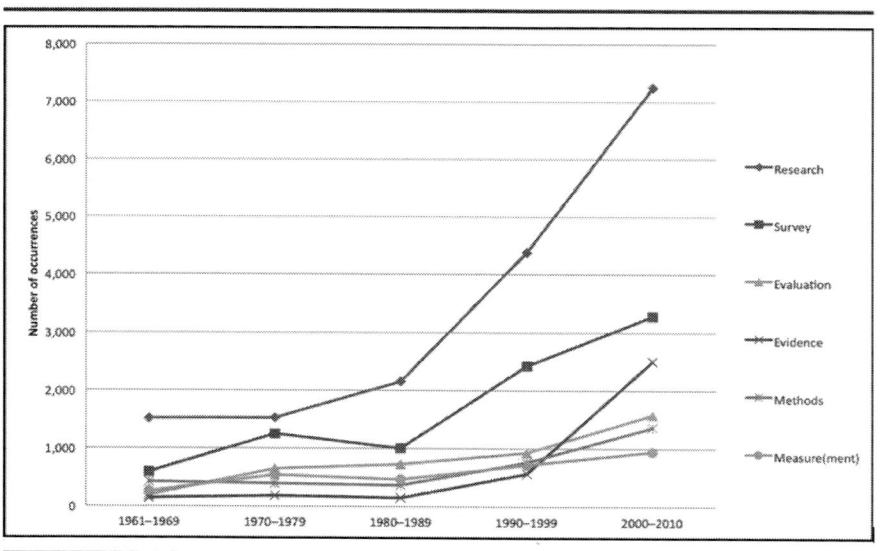

* Total occurrences of all terms and variants between 1961 and 2010539,211.

Figure 1 Number of term uses by decade*

These results provide additional data showing the rise of research and evidence-based practice in the words used by authors in the *BMLA* and *JMLA*. The data also illustrate the usefulness of the Funk corpus for investigating a variety of trends in our profession in more depth. The Funk dataset is available as an online only supplement to Funk's 2012 Doe lecture in the *JMLA* [18].

MLA research activities

A report on MLA's research initiatives from 1999– 2010 provided by MLA Executive Director Carla J. Funk, CAE, shows a lot of activity [19]. The initiatives include surveys undertaken by MLA headquarters on topics such as member salaries and competencies, benchmarking, membership, publishing, health infor- mation literacy, and consumer health information. In some instances, the research and data collection were undertaken by MLA headquarters in partnership with other organizations, such as the Pew Internet &

These results provide additional data showing the rise of research and evidence-based practice in the words used by authors in the *BMLA* and *JMLA*. The data also illustrate the usefulness of the Funk corpus for investigating a variety of trends in our profession in more depth. The Funk dataset is available as an online only supplement to Funk's 2012 Doe lecture in the *JMLA* [18].

American Life Project and the National Library of Medicine. Some research activities, such as the annual meeting evaluation, are conducted on a regular basis, while more comprehensive member surveys and *JMLA* readership surveys are periodic. MLA research activities reveal an approach that emphasizes both ongoing needs for continuous data collection in some areas and quick responses to topics of current interest, such as information specialists in context, social networking software, Magnet hospitals, personal health records, health literacy, and scholarly publish- ing, Chapters and sections of MLA are frequently involved in their own additional research and data collection efforts. The association has nine grants, scholarships, and awards that support research, including the Donald A. B. Lindberg Research Fellowship. Additional award opportunities are made available by sections and chapters.

The MLA website features research prominently on its home page. From the Research tab, it is possible to get a good overview of research activities, including the association's peer-reviewed research journal, *JMLA*. The contents of the *JMLA* as well as MLA meetings have become more research oriented with the advent of structured abstracts, an emphasis on articles that contain empirical data, and inclusion of supporting documents and data as online only appendixes. The Member Center of the website also includes the Center of Research and Education (CORE), which brings together a variety of resources for research, teaching, and learning. Of particular note are the activities of the MLA Research Section, which has published its own journal, called *Hypothesis*, since 1987. This open source journal is indexed in CINAHL and has three issues a year.

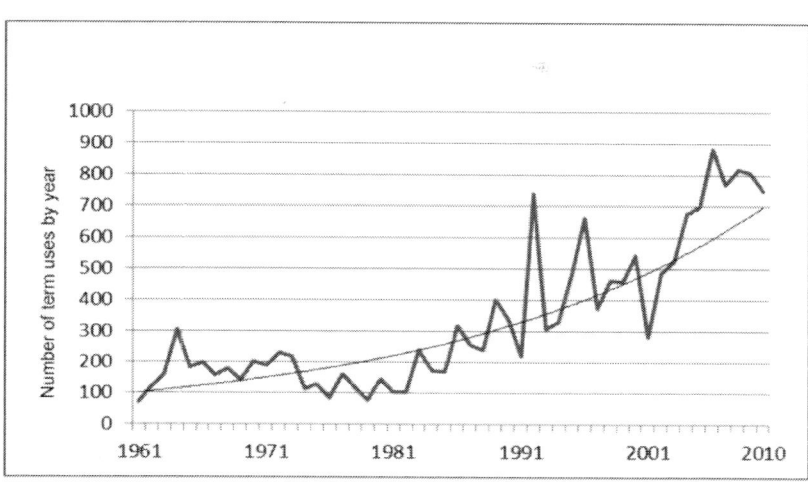

* Number of times the term "Research" or one of 88 variants appeared in a given year

Figure 2 Research*

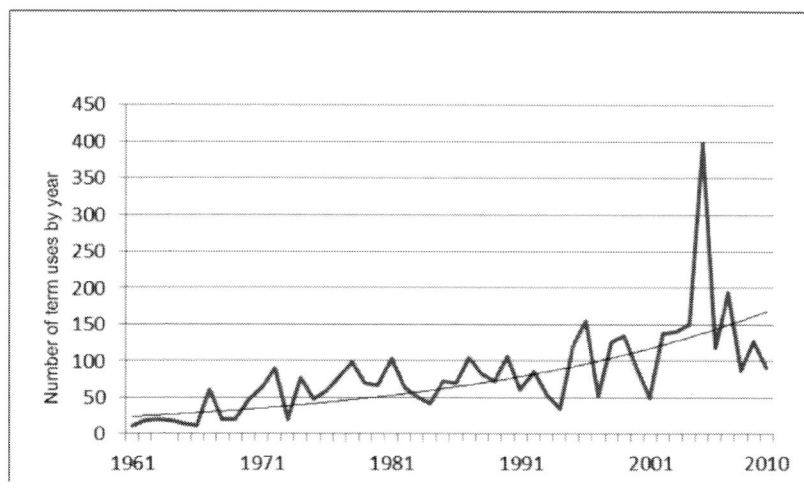

* Number of times the term "Evaluation" or one of 18 variants appeared in a given year.

Figure 3 Evaluation*

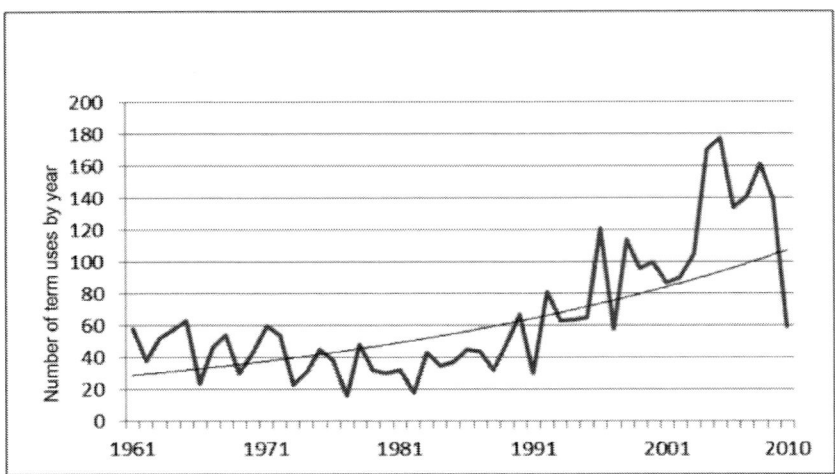

* Number of times the term "Methods" or one of 6 variants appeared in a given year.

Figure 4 Methods*

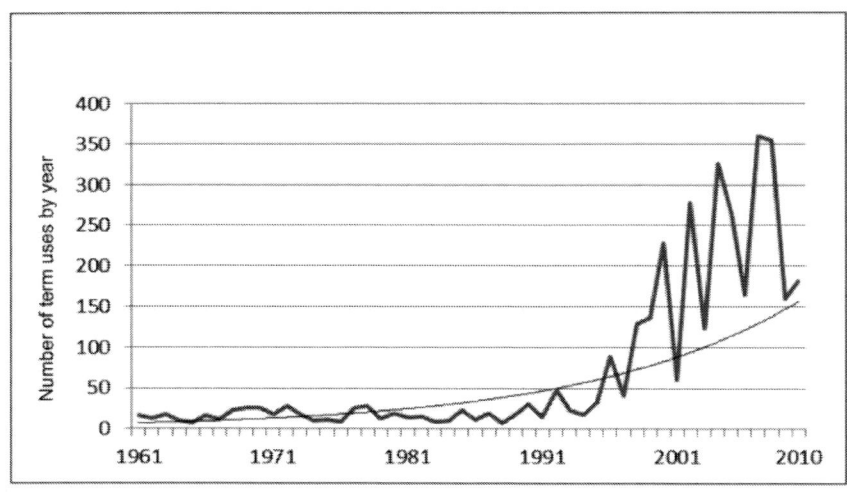

* Number of times the term "Evidence" or one of 19 variants appeared in a given year.

Figure 5 Evidence*

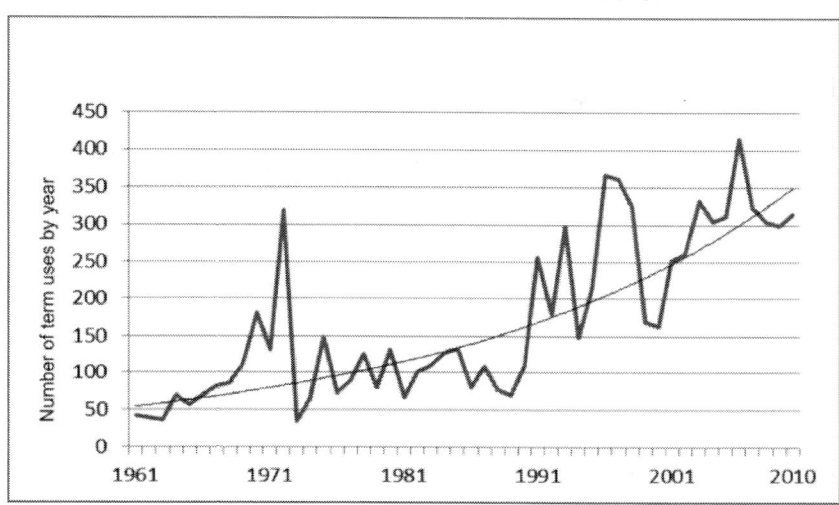

* Number of times the term "Survey" or one of 32 variants appeared in a given year

Figure 6 Survey*

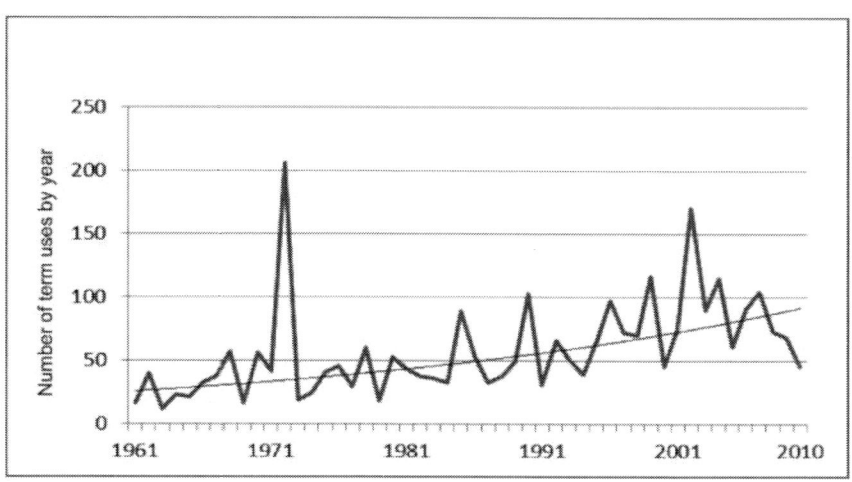

* Number of times the term "Measure(ment)" or one of 13 variants appeared in a given year.

Figure 7 Measure(ment)*

Another highly influential MLA document is the research policy statement. The original research policy issued in 1995 was entitled, "Using Scientific Evidence to Inform Information Practice." The policy describes a variety of roles for the health sciences librarian, noting the broadening of the research roles of librarians beyond giving research support to library users to becoming more active users of library and information science research, doing their own research, and applying the results of research to practice. A second edition of the research policy statement, *The Research Imperative,* appeared in 2007. It enlarges upon the original document by using video vignettes to high- light topics such as creating a culture of research, improving our knowledgebase, and exploring our research domains and the required research skills. The document also includes a list of research mile- stones 1995 to 2007 that illustrates the progress that has been made.

The MLA educational policy statement, *Competen- cies for Lifelong Learning and Professional Success,* was also revised in 2007 and includes "Research, analysis, and interpretation" as one of seven "Essential areas of knowledge" that the health sciences librarian re- quires. These areas of knowledge have been incorpo- rated into the MLA peer-reviewed professional development and career recognition program known as the Academy of Health Information Professionals. They are also used to indicate content areas in the MLA continuing education program.

Evidence-based library and information practice as an international movement

The fourth source of evidence for the rise of evidence- based practice in health sciences librarianship that I explored was at a global level. This seemed especially appropriate given the 2013 joint meeting with ICML and other international groups, and it provided an oppor- tunity to explore the international aspect of the evidence-based practice movement since its beginnings. In 2012, Jonathan Eldredge, AHIP, noted that, "Some of the most robust early EBLIP work originated in countries such as Australia, Canada, Sweden, the UK and the US" [20]. I would add Denmark and the work of Birger Hjørland [21] and his colleagues to that list. Eldredge also noted emerging interest in Japan and Iran.

This global interest has been reinforced by the "International Conference on Evidence-based Library and Information Practice" that has been held every two years since 2001. Andrew Booth from the United Kingdom was instrumental in setting up the early conferences. Booth and his colleague Ann Brice also edited the first handbook on EBLIP in 2004 [22], which included a prehistory of the movement in chapter 3. Booth was a very articulate spokesperson for the fledgling EBLIP movement in the United Kingdom, leading a series of articles with Margaret Haines in the *Library Association Record* (UK) that began in 1998 [23], three years before the first conference. The seventh "International Conference on Evidence-Based Library and Information Practice" held in 2013 in Canada had an international advisory group representing fourteen different countries.

Another major development was the establishment of the open source journal, *Evidence-Based Library and Information Practice* ,http://ejournals.library ualberta.ca/index.php/EBLIP/., at the University of Alberta in Canada in 2006. As noted on its website, the journal seeks to "provide a forum for librarians and other information professionals to discover research that may contribute to decision making in professional practice." The authors from different countries who contribute to the journal and the editorial team, with members from eleven different countries, are evidence of the international reach of evidence-based practice in our field.

As part of my exploration for this lecture, I contacted colleagues from various countries who provided me with additional detail of their activities. I also contacted or explored the work of some of the international EBLIP pioneers such as Kathleen Ann McKibbon, FMLA, Liz Bayley, Denise Koufogianna- kis, Lorie Kloda, AHIP, Virginia Wilson, Lindsay Glynn, and Jessie L. McGowan, AHIP, in Canada; Andrew Booth, Ann Brice, Margaret Haines, Christine Urquhart, Alison Brettle, Maria Grant, and others in the United Kingdom; Helen Partridge and others in Australia; Lotta Haglund, Malin Oglund, and David Herron in Sweden; and Yukiko Sakai, AHIP, in Japan. This exploration assured me that EBLIP was not only an ongoing international movement, but that it was also spreading beyond health sciences into other branches of the library and information profession.

WHERE ARE WE GOING?

Whereas traditional research has been generally been thought of as a series of steps to collect and analyze information that will increase our understanding of a topic, evidence-based practice in both medicine and librarianship has a more specific goal of improving the decision-making ability of practicing professionals. In 2012, Eldredge published a revised definition of EBLIP that combined elements from several earlier definitions. He stated that:

EBLIP provides a sequential, structured process for integrat- ing the best available evidence into making important decisions. The practitioner applies this decision making process by using the best available evidence while informed by a pragmatic perspective developed from working in the field, critical thinking skills, and an awareness of different research designs, which is further modulated by knowledge of the affected user population's values and preferences. [20]

This definition suggests a broadening of the meaning of evidence and increasing recognition of the impor- tance of professional experience and practice setting in EBLIP. The steps in evidence-based librarianship have traditionally been similar to those in evidence-based medicine: formulating an answerable research ques- tion, searching for evidence in the research literature, critically appraising the evidence found, making a decision and applying it, and evaluating outcomes. The hierarchy of desirable research methods has also been similar to that found in evidence-based medicine.

As EBLIP has evolved, librarians have found that there are many differences between medicine and librarianship that make it difficult to apply the original evidence-based practice model. Our accumu- lated research knowledgebase and literature are far more limited than that of medicine, and the type of research questions that we ask as library and information professionals are usually very different from those of practicing physicians. As a result, the hierarchy of research methods from evidence-based medicine is not always a good fit. Despite these limitations, the sources of evidence examined for this lecture show a rise in research and evidence-based practice in the profession. How can this be explained? The answer to this question lies in the creativity that has been displayed by librarians themselves as they have adapted the evidence-based practice model to fit their own needs and circumstances.

In response to some of his dissatisfaction with the original model, Booth suggested a revised model of EBLIP in 2009 [24]. Booth reviewed some of the limitations of the original five-stage model of evi- dence-based practice, noting that much decision making in librarianship is done in groups, rather than by individuals. The restrictive view of what counts as evidence may also run counter to the needs and practices of librarians. There has also been limited recognition of the complexity of decision making in librarianship. Booth outlined a new series of steps based on articulating the problem in broader terms; assembling the evidence base from multiple sources, not just the published literature; assessing the evidence through group discussion; agreeing on actions; and adapting the implementation.

This "five As" model was explored more fully in a dissertation by Koufogiannakis on how academic librarians use information in decision making [25]. Koufogiannakis found that the librarians she studied did use evidence, but not in the way that was described in the traditional evidence-based practice model. Her results pointed to the need to include local sources of evidence that take into account the context and setting in which decisions are made. Koufogiannakis also found considerable use of evidence for influencing and convincing in group decision making situations.

These findings suggest that EBLIP is evolving and changing and that we are developing our own unique approach based on our own settings and circum- stances. These new approaches provide a positive direction for the future as health sciences librarians continue to seek the most effective ways of providing quality information for improved health care by linking research to practice. In addition, there are exciting new examples of how research results are being linked to practice, such as the *JAMA* article on the value of libraries and librarians in health care by Sollenberger and Holloway [26].

ACKNOWLEDGMENTS

The author thanks all of those colleagues who have shared their knowledge and experience of evidence- based library and information practice as well as Marlys Ray for her assistance in the preparation of the manuscript for publication.

REFERENCES

1. Doe J. The survey and after. Bull Med Lib Assoc. 1961 Jul;49(3):361–8.

2. Marshall JG, Neufeld VR. A randomized trial of librarian educational participation in clinical settings. J Med Educ. 1981 May;56(5):409–16.

3. Department of Clinical Epidemiology and Biostatistics, McMaster University. How to read clinical journals, I: why to read them and how to start reading them critically. Can Med Assoc J. 1981 Mar 1;124(5):555–8.

4. Guyatt GH, Rennie D. Users' guides to the medical literature. JAMA. 1993 Nov 3;270(17):2096.

5. Sackett DL, Haynes RB, Guyatt GH, Tugwell P. Clinical epidemiology: a basic science for clinical medicine. 2nd ed. Boston, MA: Little Brown & Co; 1991.

6. Sackett DL, Rosenberg WMC, Gray JAM, Haynes RB, Richardson WS. Evidence-based medicine: what it is and what it isn't. BMJ. 1996 Jan 13;312(7023):71.

7. Meyerhoff E. Foundations of medical librarianship. Bull Med Lib Assoc. 1977 Oct;65(4):409–18.

8. McClure LW. The promise of fruit…and light. Bull Med Lib Assoc. 1985 Oct;73(4):319–29.

9. Brodman E. Choosing physiology journals. Bull Med Lib Assoc. 1944 Oct;32(4):479–83.

10. Holtz VH. Measures of excellence: the search for the gold standard. Bull Med Lib Assoc. 1986 Oct;74(4):305–14.

11. Love E. The science of librarianship: investing in the future. Bull Med Lib Assoc. 1987 Oct;75(4):302–9.

12. Anderson RK. Reinventing the medical librarian. Bull Med Lib Assoc. 1989 Oct;77(4):323–31.

13. Matheson NW. The idea of the library in the twenty-first century. Bull Med Lib Assoc. 1995 Jan;83(1):1–7.

14. Peay WJ. Strategies and measures for our next century. Bull Med Lib Assoc. 1999 Jan;87(1):1–8.

15. Fuller SS. Enabling, empowering, inspiring: research and mentorship through the years. Bull Med Lib Assoc. 2000 Jan;88(1):1–10.

16. Roper F. The Medical Library Association's professional development program: a look back at the way ahead. J Med Lib Assoc. 2006 Jan;94(1):8–18.

17. Plutchak TS. Breaking the barriers of time and space: the dawning of the great age of librarians. J Med Lib Assoc. 2012 Jan;100(1):10–9..

18. Funk ME. Our words, our story: a textual analysis of articles published in the Bulletin of the Medical Library Association/Journal of the Medical Library Association from 1961 to 2010. J Med Lib Assoc. 2013 Jan;101(1):12–20.

19. Funk C. Summary of research initiatives by the Medical Library Association, 1999/2000–2009/2010 [unpublished document]. Chicago, IL: Medical Library Association.

20. Eldredge JD. The evolution of evidence-based library and information practice, part 1: defining EBLIP. Evid Based Lib Inf Pract. 2012;7(4):139–45.

21. Hjørland B. Evidence-based practice: an analysis based on the philosophy of science. J Am Soc Inf Sci. 2011 Jul;62(7):1301–10.

22. Booth A, Brice A, eds. Evidence-based practice: a handbook for information professionals. London, UK: Facet; 2004.

23. Booth A, Haines M. Room for a view? Lib Assoc Rec. 1998;100(8):411–2.

24. Booth A. EBLIP five-point-zero: towards a collaborative model of evidence-based practice. Health Info Lib J. 2009 Dec;26:341–44. Koufogiannakis DA. How academic librarians use evidence in their decision making: reconsidering the evidence-based model [PhD dissertation]. Aberystwyth, UK: Department of Information Studies, Aberystwyth University; 2013.

25. Sollenberger JF, Holloway RG. The evolving role and value of libraries and librarians in health care. JAMA. 2013 Sep 25;310(12):1231–2.

26. Sollenberger JF, Holloway RG. The evolving role and value of libraries and librarians in health care. JAMA. 2013 Sep 25;310(12):1231–2.

5

CHAPTER

CURRENT PRACTICES IN LIBRARY/INFORMATICS INSTRUCTION IN ACADEMIC LIBRARIES SERVING MEDICAL SCHOOLS IN THE WESTERN UNITED STATES: A THREE-PHASE ACTION RESEARCH STUDY

Jonathan D Eldredge1, Karen M Heskett2, Terry Henner3 and Josephine P Tan4

[1]Health Sciences Library & Informatics Center and Department of Family & Community Medicine, University of New Mexico, MSC09 5100, Albuquerque, NM 87131-0001, USA

[2]UC San Diego Biomedical Library, UC San Diego, 9500 Gilman Dr. 0699, La Jolla, CA 92093, USA

[3]Savitt Medical Library, University of Nevada School of Medicine, Reno NV 89557, USA

[4]UCSF Library and Center for Knowledge Management, UCSF, 530 Parnassus Avenue, San Francisco, CA 94143-0840, USA

ABSTRACT

Background

To conduct a systematic assessment of library and informatics training at accredited Western U.S. medical schools. To provide a structured description of core practices, detect trends through comparisons across institutions, and to identify innovative training approaches at the medical schools.

Methods

Action research study pursued through three phases. The first phase used inductive analysis on reported library and informatics skills training via publicly-facing websites at accredited medical schools and the academic health sciences libraries serving those medical schools. Phase Two consisted of a survey of the librarians who provide this training to undergraduate medical education students at the Western U.S. medical schools. The survey revealed gaps in forming a complete picture of current practices, thereby generating additional questions that were answered through the Phase Three in-depth interviews.

Results

Publicly-facing websites reviewed in Phase One offered uneven information about library and informatics training at Western U.S. medical schools. The Phase Two survey resulted in a 77% response rate. The survey produced a clearer picture of current practices of library and informatics training. The survey also determined the readiness of medical students to pass certain aspects of the United

States Medical Licensure Exam. Most librarians interacted with medical school curricular leaders through either curricula committees or through individual contacts. Librarians averaged three (3) interventions for training within the four-year curricula with greatest emphasis upon the first and third years. Library/informatics training was integrated fully into the respective curricula in almost all cases. Most training involved active learning approaches, specifically within Problem-Based Learning or Evidence-Based Medicine contexts. The Phase Three interviews revealed that librarians are engaged with the medical schools' curricular leaders, they are respected for their knowledge and teaching skills, and that they need to continually adapt to changes in curricula.

Conclusions

This study offers a long overdue, systematic view of current practices of library/informatics training at Western U.S. medical schools. Medical educators, particularly curricular leaders, will find opportunities in this study's results for more productive collaborations with the librarians responsible for library and informatics training at their medical schools.

KEYWORDS

Medical libraries, Medical informatics, Teaching, Active learning, Curriculum, Library science, Information science, Information literacy, Information fluency, Information seeking behavior

1. BACKGROUND

Medical students must master skills to retrieve, critically assess, and integrate biomedical information into their clinical decision-making. These skills are recognized as core competencies. As Golub has noted, "The relatively short half-life of medical knowledge has led to the recognition of the importance of instilling the value and the skills of life-long learning as a core piece of medical education" [1]. Accordingly, over the past 75 years academic health sciences librarians have delivered information skills training as part of the formal education of medical students. William Dosité Postell, reporting on a survey conducted during the 1930s, indicated that 50 of the 64 medical schools in the U.S. (78%) offered library instruction [2]. Earl's 1996 report on a survey of 123 academic health sciences libraries produced 55 responses with 75% reporting that they provided library instruction to medical students [3].

The 1982 Matheson Report advised educators that medical education in the future would bear little resemblance to the past due to a daunting expansion of medical information. Future physicians, while still in medical school, would need to acquire a new set of skills to manage and interpret the huge volume of information [4]. The Association of American Medical Colleges' (AAMC) inventory of informatics competencies prompted some academic health sciences libraries in the U.S. to reassess, revamp, and redeploy their library instruction programs to better prepare medical students for a future requiring sophisticated information seeking skills. The arrival of these AAMC competencies generated a great deal of discussion among health sciences librarians, but it remained unclear as to the extent that librarians were ensuring that these AAMC competencies were integrated into medical school curricula [5,6].

Health sciences librarians perform a variety of expected and unexpected roles in U.S. medical school curricula, as validated by an extensive review of studies [7]. Health sciences librarians in the western U.S. have reported on a number of studies that focus on novel or effective library instruction approaches to training medical students at individual academic health sciences libraries [8-32]. No recent surveys have updated Earl's 1996 study, however; and, there is an absence of research that reports comprehensively on the state of library instruction in the western region of the U.S.

Concerns about these research gaps drew the interest of a regional chapter of Libraries in Medical Education (LiME), an interest group of the Association of American Medical Colleges (AAMC) Group on Educational Affairs. LiME/AAMC meets annually as a means for members to report on current instruction related activities of librarians at institutions in the region. Wishing to take a more systematic and comprehensive approach, in 2009 a LiME research task group undertook an environmental scan of library instruction for medical students at all academic health sciences libraries serving accredited medical schools in the Western United States. The long term goal of the task force was to create a group of interested participants who could support a process of data gathering and reflection on current practices in order to improve the integration of library instruction into medical education. The purpose of this study was to facilitate broad comparisons between peer libraries by exploring in a comprehensive and systematic manner the ways in which academic health sciences libraries in the Western United States deliver instruction to medical students.

2. METHODS

The investigators implemented a three phase action research project consisting of (1) a descriptive environmental scan, (2) survey, and (3) interview methodologies. The present study included the common action research elements of researcher participation, real-life field settings, and reflective periods [33]. Vezzosi's use of an action research approach to understand the effectiveness of library instruction represents a model of how action research can be employed in this subject area [34]. Somekh delineates eight principles normally found in action research in education contexts. The present study incorporated seven of those principles: a cyclical process; collaborative partnerships; knowledge development; roles of the researchers in the process; exploratory engagement; researchers as learners; and a broad contextual awareness [35].

Phase one

Guided by discussion at LiME meetings and conversations between task force members, Phase One consisted of an unobtrusive environmental scan of publicly facing websites of academic health sciences libraries and educational institutions they serve, focusing on the 17 accredited medical schools of the Association of American Medical Colleges (AAMC) in the Western U.S. listed in Table 1. The investigators sought to construct a detailed picture of educational activities conducted by medical librarians and to identify common patterns of curricular support. Team members made preliminary investigations of public-facing websites at the institutions in the western U.S. Through an iterative process of review, reflection, synthesis, and discussion team members devised a checklist to apply to all 17 sites. This team-generated checklist guided reviewers in examining publicly-available documents such as library newsletters, course guides, and annual reports as well as relevant data from the Association of Academic Health Sciences Libraries (AAHSL) [36]. During the process, the investigators looked for unique or innovative library instruction practices. They also identified basic descriptive information about the user population of the library and, in some cases, information about the faculty status and committee appointments of library staff.

Despite the variable quality and quantity of the initial results, Phase One provided useful information to help investigators articulate the following three research questions to guide phases two and three:

1. What are the current core or commonly followed practices of teaching library/informatics skills to medical students?

2. What patterns or possible trends might emerge from comparisons of different academic health sciences libraries in the Western US that provide library/informatics skills trainings for medical students?

3. What innovative practices can be identified at specific academic health sciences libraries that might be adapted to other academic health sciences libraries?

Table 1 Potential and actual participants in phases 1 & 2: academic libraries supporting schools of medicine

University & Library	Responded to phase 2
1. Charles Drew University of Medicine & Science, Health Sciences Library	✓
2. Loma Linda University Medical Center, Jesse Medical Library & Information Center	✓
3. Oregon Health and Science University, Library	
4. Stanford University Medical Center, Lane Medical Library	
5. University of Arizona (Tucson Campus), Arizona Health Sciences Library	✓
(University of Arizona (Phoenix Campus) Partnership of U of A & ASU medical school dissolved mid-project. ASU counted as part of U of A)	
6. University of California, Davis, Carlson Health Sciences Library	✓
7. University of California, Irvine, Science Library	✓
8. University of California, Los Angeles, Biomedical Library	
(University of California, Riverside program is developing, with most services provided by UCLA and therefore, counted under UCLA)	
9. University of California San Diego, Biomedical Library	✓
10. University of California, San Francisco, Library	✓
11. University of Colorado, Health Sciences Library	✓
12. University of Hawaii at Manoa, Health Sciences Library	
13. University of Nevada Reno, Savitt Medical Library	✓
14. University of New Mexico School of Medicine, Health Sciences Library and Informatics	✓
15. University of Southern California, Norris Medical Library	✓
16. University of Utah, Eccles Health Sciences Library	✓
17. University of Washington, Health Sciences Library	✓

Phase two

The team shared its analysis of the Phase one data with the larger LiME membership for comment and discussion to guide the design and distribution of a descriptive survey [37]. The survey's final format incorporated the Phase One unobtrusive study data, the investigators' own library instruction experiences, feedback from the (AAMC/LiME) group, and anecdotal knowledge of instructional activities typical in health sciences libraries.

The investigators designed the survey to learn: the medical school governance structure, the role (if any) of librarians in that governance structure, details about library instruction integrated within the curriculum, library instruction (if any) not integrated within the curriculum, faculty status, how library/informatics instruction skills were assessed, and a prediction as to whether graduating medical students at their institution would perform well on PubMed database searches on a United States Medical Licensure Exam (USMLE) currently under consideration by the National Board of Medical Examiners [38,39]. Additional file 1 contains the Phase Two survey questions.

The investigators secured Institutional Review Board approval (HRPO # 10–102) from the University of New Mexico to conduct the survey and any follow-up interviews in the last phase. The investigators deployed the invitation to complete the survey on April 7, 2010. The investigators emailed this invitation to the directors of all 17 academic health sciences libraries serving accredited medical schools in the Western U.S. as listed in the AAHSL Directory [40]. The directors were asked to forward the emailed invitation to all library employees responsible for conducting library instruction with medical students, as a modified form of snowball sampling. A total of three reminder emails were sent and the final invitation was sent in mid-June with an announced closing date of June 22, 2010. The invitation required all respondents to consent to participate in accordance with ethical research principles and invitees were asked to click on a link to the survey as their means of giving consent. Table 1 lists the institutions contacted with checkmarks aside those institutions responding to the survey. The investigators compiled the survey responses, discussed them at length via online conferencing software, and synthesized the data. In keeping with the reflective phase of action research, the results were shared with the librarian community in a panel presentation at a regional meeting of WGEA [41]. The ensuing commentary and discussion among meeting attendees were critical in devising the third phase of the study.

Phase three

This phase of the project consisted of the investigators developing and deploying a standardized template of six (6) interview questions. The template additionally included some prompts intended to follow these specific questions so the interviewer might pursue any productive avenues for further discussion. The investigators interviewed the respondents at each institution who had the greatest breadth and depth of library instruction experience with medical students. The structured interview questions, and the prompts for possible follow-up, appear in Additional file 2. The investigators implemented the follow-up interviews lasting approximately 30 minutes each by telephone or online conference software beginning in December 2010 and completed the structured interviews during April 2011. All interviewees were sent summaries of the interviews so they might correct any responses, or add clarifying text.

3. RESULTS

This three-phase action research study produced results on the state of library/informatics training that can both inform current practices for medical educators and point toward future research. The environmental scan in Phase One generated targeted research questions about current practices while Phases Two and Three predominantly painted a picture of current practices.

Phase one results

The information gathered from the websites ranged from ones that merely outlined the essential library services offered extending all the way to websites offering comprehensive accreditation self-study reports in accordance with the standards set by the Liaison Committee on Medical Education guidelines [42]. Inspection of the institutional websites revealed announcements of upcoming workshops, links to handouts from educational sessions and workshops, indications of curriculum-based courses, links to online multimedia tutorials, and access to supplementary instructional guides developed by librarians. While some of the institutions' websites provided a complete picture of their library instruction activities, many lacked sufficient detail to accurately portray the roles that librarians play in supporting medical school curricula. The investigators recognize that some of this information might have been behind password protected websites and thereby unavailable. As noted earlier, the constraints of this purely descriptive approach resulted in an incomplete and inconclusive picture of library instructional programs. An analysis of gaps in the data helped to shape subsequent phases of the study and enabled investigators to generate targeted survey questions intended to yield comparative information about library instruction to medical students.

Phase two results

Colleagues at 13 of 17 eligible academic health sciences libraries completed the survey, a response rate of 77%. Two librarians from one library completed the survey, and as their responses were consistent with one another, the investigators merged these responses. An informal follow up by one investigator with colleagues at three of the four non-responding libraries revealed that they did not have time to complete the survey. No significant geographic, governance, or other recognizable characteristics distinguished the non-responders from those who responded to the survey. For the responding libraries, all 13 medical schools governed their curricula with a curriculum committee. Academic health sciences librarians interacted with these curriculum committees directly through a variety of methods including regular membership, ex-officio membership, specialized subordinate groups, regular meetings with curricular leaders, or via informal contacts. The plurality of responses indicated that most organizations had multiple means of interaction but the primary method was via ex-officio membership on curriculum committees. A little more than half (53%) of the respondents had faculty status at their respective library and one also had an academic appointment through the school of medicine. All others had academic promotion systems equivalent to faculty status within their institutions. The respondents had an average of 18.4 years of experience as librarians and only three respondents had fewer than 10 years of experience. The majority of respondents had been involved in the most recent Liaison Committee on Medical Education accreditation review process.

The Phase Two survey emphasized identifying instances where librarians engaged in curricular-based library interactions with medical students. All but one of the 13 institutions required incoming medical students to attend basic library orientation sessions. In total, 53 discrete sessions were described along with the year in which the students experienced the sessions. Responses showed activity occurred across the undergraduate curriculum. In general, librarians had an average of three (3) interventions integrated within the core curricula. Not surprising, first year medical students were the target audience for the majority of sessions (29 in total). Third year medical students were the second most frequently contacted audience (21 sessions) followed closely by the second year students (with 16 sessions). See Figure 1. Some sessions were composed of a mix of students from different years. The majority of sessions ($n = 44$) were required with less than 20% ($n = 9$ sessions) as elective sessions]. Fourth year student activity consisted primarily of liaison contacts or consults.

Descriptions of the instruction sessions by respondents were consistent across the different institutions so that, even with institutional variances, responses could be categorized and quantified. Figure 2 summarizes the five (5) types of instruction sessions that emerged: hands-on, lecture, virtual, non-specific orientation, and required consults. Hands-on sessions included anything described with that term or a description indicating student interactions or student practice. Lecture sessions include those described as such as well as ones described as multiple week sessions. Hands-on sessions and lecture sessions were indicated equally with 19 sessions each. Virtual instruction is a growing trend in libraries [43] and the librarian medical educators noted 8 virtual instruction sessions which included work through blogs, online student peer assessment, wikis, videos, or online tutorials. Orientation sessions, not otherwise described, were left as such and termed non-specific orientation. See Table 2.

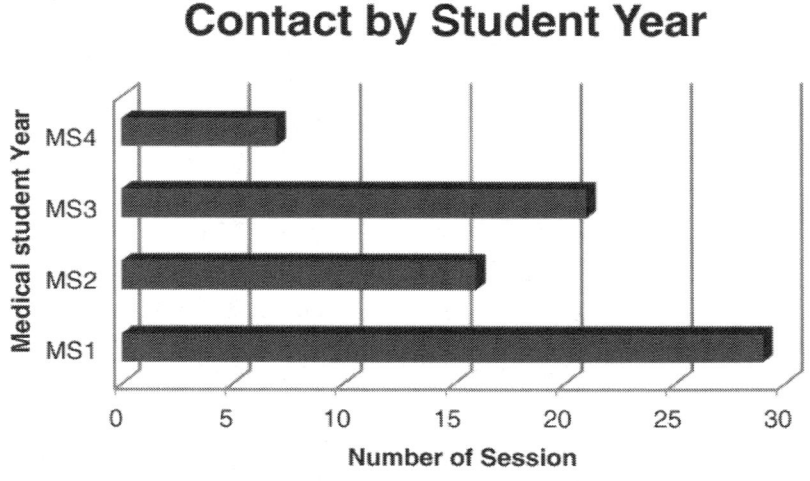

Figure 1. Library instruction by medical student year.

Type of Instruction Session

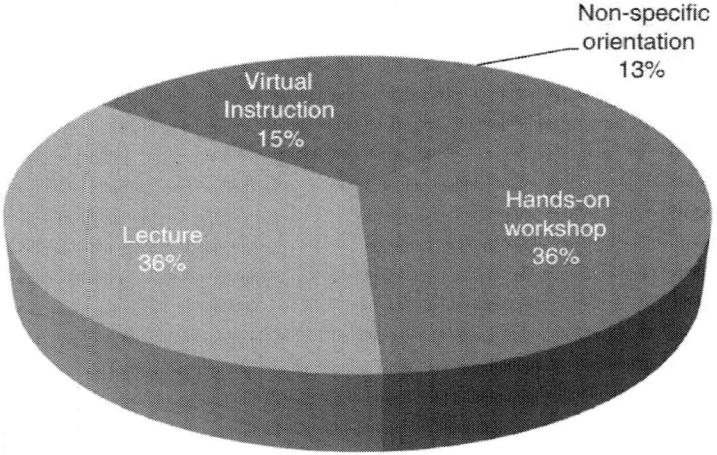

Figure 2. Types of librarian instruction.

Table 2 Number of instruction sessions by format

Format	Total
Hands-on workshop	19
Lecture	19
Virtual Instruction	8
Non-specific orientation	7
Required Consult	1

Other than ubiquitous PubMed sessions, two distinct topics were volunteered in the descriptions – evidence-based medicine (EBM) (23 sessions) and problem-based learning instruction (5 sessions). Figure 3 indicates that faculty status does not appear to have an impact on curriculum-integrated session *except* that faculty librarians tend to offer a few more required sessions (i.e., fourth and fifth sessions). Only a couple of the non-faculty librarians offered more than three sessions, and these were not always required. One-quarter of the descriptions voluntarily detailed time spent on instruction activities and future iterations of this study might request this specific information. For this small subset, the average time spent on instruction was 2 hours – ranging from a minimum of 30 minutes to a maximum of 32 hours (for a multi-week sequence).

Nearly one-quarter of the libraries reported an assessment of medical students prior to instruction. Formal assessment within the curriculum seems to be a rarely performed activity for librarians, however. As part of the curriculum, schools of medicine have some assessment of knowledge and skills, but it is unclear how the librarians are involved with that activity. Those responding to this question, representing 5 of the 13 schools, submitted just 10 instances of assessment. Yet, as a comparison, over 50 instruction sessions were entered in the survey. Of the assessments, a total of 9 were graded or pass/fail assignments with 2 having a self-assessment or peer-assessment component. Most of the described assessments involved activities such as finding resources and evaluating search skills in order to answer questions. Four sessions dealt specifically with evidence-based medicine (EBM) topics and only two specifically mentioned dealing with citations. The themes identified in Phase Two survey results are consistent with the literature in suggesting that medical students have a diminished preference for non-specific library orientations that lack a curricular context and focused learning objectives.

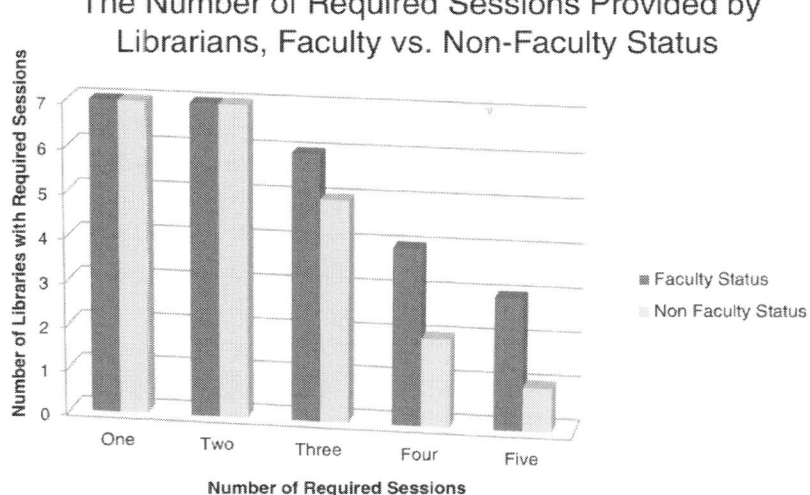

Figure 3. Number of instruction sessions by librarian faculty vs. librarian academics, non-faculty status. Faculty Status, Non-faculty status.

Phase three results

During the autumn of 2010, the investigators held several in-depth online conference meetings to discuss the survey results. The structured interview questions to be used in Phase Three emerged from this action research process of review, reflection, and discussion. Twelve (12) of the 13 survey respondents were able to arrange interviews with an investigator during the allotted timeframe, a participation rate of 92%. The 12 interviews occurred during the December 2010 to April 2011 time period. The investigators engaged in both synchronous and asynchronous discussions to reach consensus on their interpretations of these interviews as summarized underneath each of the following italicized questions.

1. Could you explain the reasons for the successes you have experienced in integrating information literacy/fluency/competencies into your medical school's curriculum?

Answers varied widely, but some recurring themes emerged from the combined interviews:

- Librarians are engaged with the medical school curriculum committee and with curricular leaders.

- Librarians' efforts frequently rely upon "champions" within the medical school who can advocate for integrating library/informatics skills.

- Librarians have strong support for library or informatics instruction from the library administration.

- Librarians have proven themselves to their teaching faculty colleagues or medical school administrators over time by demonstrating both their knowledge and teaching skills.

2. If we created a supplement to our upcoming article in a publicly accessible institutional repository that contains samples of outstanding handouts or other documents, would you be willing to contribute 3–5 of your best items?

- Responses point to a willingness to donate instruction related materials as well as enthusiasm for creating an open access archive.

3. Could you describe your online curricular or instructional support (examples: learning management system such as Blackboard; social networking; chat) at your institution? Does the library or another unit such as IT provide this support?

Most medical schools use a commercial learning management system. Most also use a locally produced learning management system to supplement the commercial system in order to meet all of their needs.

4. What were the "lessons learned" from past mistakes or miscalculations in your efforts?

- Free-standing courses never work as well as library instruction that is integrated fully into the curriculum

- The need to keep adapting to changing circumstances, including curricular changes, in the medical school

- Secure detailed feedback from students on the quality of teaching, its relevance to curricular content, and the content taught

- Perseverance despite setbacks usually leads to success

5. Why are librarians at your library motivated to teach?

Most librarians taught because of their faculty status, or were expected to teach due to a similar codified equivalent of faculty status as an institutional career ladder for promotion. Beyond this broad expectation, however, respondents noted that most librarians teach as a natural outgrowth of their desire to ensure that medical students (and future physicians) possess all needed library/informatics skills. One respondent mentioned that there were too few librarians to teach these skills on an individual point-of-use basis so formal instruction was the only reasonable cost-effective option. Interestingly, multiple authors made this central cost-effectiveness argument in a classic volume published in 1974 during a renaissance within library instruction in academic libraries [44]. Additionally, most respondents indicated that those librarians who teach certainly enjoy this instructional role.

6. Reviewing your responses concerning your activities, how much time was devoted to each?

Respondents' formal work allocation to the education of medical students encompassed anywhere on average from 15 to 50% of their overall efforts. Most respondents reported that they spend a considerable amount of time outside of the classroom with curricular design, keeping abreast of curricular changes, and preparing to teach. On this latter point, one respondent mentioned spending 25 hours to perfect a presentation for a single one hour session in front of medical students since she realized that her time was so limited within today's "crowded curriculum" [45] at US medical schools.

Figure 4 provides a Wordle™ word cloud that visually displays the words used most frequently by interview respondents. The investigators expected to find words such as "librarians" and "library" prominently displayed in the word cloud. The investigators did not expect to see the words "teaching", "medical students", "curriculum", or "faculty" so frequently mentioned. Thus, the word cloud discovered some less obvious patterns otherwise lost by reading the texts of the structured interviews compiled in Phase Three.

Figure 4. Wordle cloud from phase 3 interviews.

4. DISCUSSION

This study fills a gap in health sciences library/informatics skills training at different US medical schools. The investigators discovered that many of their colleagues achieved success in integrating library and informatics skills into their respective curricula. This project readily confirmed the diversity of practices. This study also produced suggestive non-statistical evidence for librarians' status and roles in curricular governance.

An action research approach

Consistent with the tenets of an action research approach, the investigators followed an evolutionary, developmental course in order to better explore the challenges facing library/informatics instructors in medical education. By examining and sharing data in a stepwise approach, investigators were able to integrate discussion and concerns from the practitioner community in order to improve each subsequent phase of inquiry. This iterative process of engagement contributed significantly to the intended goal of producing a useful report on practices and trends in library instruction.

Because the investigators were members of the very AAMC/LiME community of practitioners under study, they bridged conventional forms of dichotomy between themselves and their subjects, contributing to a collaborative co-construction of knowledge [33]. Incorporation of key aspects of action research in the study, including building relationships, acknowledging and sharing power, and encouraging participation of the study population [34], enhanced the eventual applicability of results to professional practice.

Phase one

This phase revealed that an institution's publicly facing website cannot be relied upon to gather enough data to make more than just superficial comparisons across institutions on library education. The investigators learned in this process moreover that the availability of more robust data would be inconsistent across institutions, at best.

Phase two

The survey addressed many of the investigators' questions generated during Phase One. A particular focus examined the extent to which curriculum integration is reflected in library instructional activities. The literature has long suggested that increased educational effectiveness and impact on student learning is predicated on integration of library instruction into the existing medical curriculum, rather than as a separate component of a library's educational program [46].

In her landmark article, Francesca Allegri defines "course integrated instruction" as having met at least three of the four following criteria: "(1) faculty outside the library are involved in the design, execution, and evaluation of the program, (2) the instruction is curriculum-based, in other words, directly related to the students' course work and/or assignments, (3) students are required to participate, and (4) the students' work is graded or credit is received for participation [47]." Survey responses reflected and met Allegri's definition of course integrated instruction. Respondents described a variety of curriculum integration activities such as recurring roles in semester long classes; collaborative teaching of informatics concepts to support problem-based learning exercises, and interactive instruction covering content tied directly to exam questions. EBM training has evolved over the past few decades with librarians having a growing role in working with both students and faculty within the curriculum [48,49]. The survey responses indicated that over 40% of the sessions were EBM topics, a finding that validates much of the research in this area.

Faculty status or its close equivalent for librarians appears to provide access and credibility for librarians needing to integrate library or informatics training into medical school curricula. Librarians and teaching faculty members alike seemed to recognize their mutual interdependence in these endeavors. One of the founders of the modern library instruction movement, Evan Farber, has emphasized this mutually-dependent relationship between librarians and their teaching faculty colleagues [50]. Travis has admonished her colleagues more recently that "Librarians need to think and act globally, never compartmentalize library instruction efforts, and find ways to scale information literacy into an institution wide model [51]." Librarians at the institutions in this study apparently were paralleling Travis' advice as further evidenced by their successes. Librarians involved in providing integrated library and informatics instruction had an average of 18 years' experience, which strongly suggests that this role requires considerable experience, knowledge, and expertise. Wiggins similarly has noted that library or informatics instruction often succeeds when the experienced and knowledgeable librarian can provide skeptical students with the rationales for the relevance of library instruction at a specific juncture in the curriculum [52]. The recent resurgence of interest among educators on the national level in cultivating affective educational objectives also dovetails with this data [53].

Phase three

The Phase Three interviews highlighted the importance of having champions among the teaching faculty and the support of administrators overseeing the curriculum. Curzon has emphasized the importance of such partnerships, particularly with teaching faculty who must balance a crowded curriculum with the student's escalating need to effectively manage the exploding information universe [54]. In the absence of a context or perceived need among the students, interview respondents reported that the basic library orientation sessions tend to have poor educational outcomes. Prior research had suggested that library/informatics instruction most likely will be more effective when integrated into the curriculum [55]. This study preceded publication of Moore's 2011 sentinel *Academic Medicine* commentary on the need for library/informatics training. The findings in this study provide supplementary evidence to support Moore's thesis [56] as well as revealed innovative ways librarians are maximizing limited instruction time with their curricular partners.

Limitations

This study details an environmental scan that explored the breadth and depth of library/informatics skills instruction for medical students at academic health sciences libraries in the Western U.S., and represents a unique examination of a largely uncharted subject area. The authors could identify only one account that bore even a distant similarity to the approach found in the present study [57]. The research reported in this present study cannot be generalized to the entire U.S. due to the geographic concentration in the western region, the small number of institutions, and the investigators' awareness of diverse library instruction practices in other regions. The survey responses also constituted low-level frequency and descriptive data that could not be easily categorized into discrete data points. Still, medical educators and librarians outside the region can benefit from learning about the rich and diverse descriptive information on how their colleagues at different western U.S. institutions grapple with challenges similar to their own. In the process of implementing this three phase action research project the investigators have created a template for a national level action research study. This template could even be modified to secure more defined responses, if viewed by colleagues elsewhere as desirable. The investigators would be happy to share with interested colleagues their experiences in conducting this type of multiple methods study.

Future research

Expanding the focus of this research beyond the Western region would provide a sufficient sample of librarians to make statistically significant test of the following hypotheses:

1. Great diversity in how medical students are trained on library/informatics skills exists in the United States, and that knowledge of some of these practices will be valued by colleagues involved in similar types of library/informatics training.

2. A correlation exists between librarian roles in governance structures *and* their degree of involvement in training medical students on library/informatics skills, the degree to which this training has been integrated into the curriculum, and their assessment of medical student performance.

5. CONCLUSION

This study provides medical educators and librarians with a detailed snapshot illustrating the current nature of library instruction in medical schools. It delineates the degree to which library/informatics competencies are integrated into medical school curricula. Analysis of the information examines some preconditions for successful instructional programs, reveals challenges shared by librarian instructors, and discusses adaptive strategies that have led to greater student satisfaction. The results reinforce the notion that information skills instruction is an important part of medical education and are indicative of the value librarians contribute to the educational process.

Medical educators, if not already doing so, should actively partner with librarians at their institution to strive for curriculum integrated information skills training of medical students. Librarians should also ensure that feedback on library instruction is included as a standard component of student course evaluations. Folding evaluations of library instruction into the broader curricular context may increase the validity of student feedback, give instructors meaningful data with which to quantify skills improvement, enhance future library instruction, and relieve students of the burden of completing separate post-instruction library surveys. Librarians play a pivotal role in providing the skills to bolster life-long learning that goes well beyond medical school and prepares a solid foundation for how to keep up with the ever-growing body of medical education research literature.

COMPETING INTERESTS

The authors declare that they have no competing interests.

AUTHORS' CONTRIBUTIONS

JE, KH, TH, and JT conceived of this project. JE developed the design and secured IRB approval and subsequent renewals. JE, KH, TH, and JT conducted the Phase One environmental scan and designed the Phase Two survey. KH implemented the survey via SurveyMonkey™

and codified and tabulated the results. JE, KH, and TH interpreted the survey results and designed the follow-up interview questions in Phase Three. JE, KH, TH, and JT conducted the in-depth interviews in Phase Three. JE, KH, TH, and JT wrote and edited the manuscript. All authors read and approved the final manuscript.

ACKNOWLEDGEMENTS

The authors wish to thank the librarians at the AAMC's LiME WGEA 2009 meeting for the inspiration for this project. Thanks also to all the librarians who provided such rich information on the survey and interviews.

AUTHOR DETAILS

1 Health Sciences Library & Informatics Center and Department of Family & Community Medicine, University of New Mexico, MSC09 5100, Albuquerque, NM 87131-0001, USA. 2 UC San Diego Biomedical Library, UC San Diego, 9500 Gilman Dr. 0699, La Jolla, CA 92093, USA. 3 Savitt Medical Library, University of Nevada School of Medicine, Reno NV 89557, USA. 4 UCSF Library and Center for Knowledge Management, UCSF, 530 Parnassus Avenue, San Francisco, CA 94143-0840, USA.

REFERENCES

1. Golub RM. Medical education theme issue 2008: call for papers. JAMA.2007;298(22):2677.

2. Postell WD. Further notes on the instruction of medical school students in medical bibliography. Bull Med Libr Assoc. 1944;32(2):217–220.

3. Earl MF. Library instruction in the medical school curriculum: a survey of medical college libraries. Bull Med Libr Assoc. 1996;84(2):191–195.

4. Matheson NW, Cooper JA. Academic information in the academic health sciences center. roles for the library in information management. J Med Educ. 1982;57(10 Pt 2):1–93.

5. Contemporary issues in medical education: medical informatics and population health.

6. McGowan JJ, Passiment M, Hoffman HM. Educating medical students as competent users of health information technologies: the MSOP data. Stud Health Technol Inform.2007;129(2):1414.

7. Schwartz DG, Blobaum PM, Shipman JP, Markwell LG, Marshall JG. The health sciences librarian in medical education: a vital pathways project task force. J Med Libr Assoc. 2009;97(4):280–284.

8. Kingsley K, Galbraith GM, Herring M, Stowers E, Stewart T, Kingsley KV. Why not just google it? An assessment of information literacy skills in a biomedical science curriculum. BMC Med Educ. 2011;11:17.

9. Geppert CM, Arndell CL, Clithero A, Dow-Velarde LA, Eldredge JP, Kalishman S, Kaufman A, McGrew MC, Snyder TM, Solan BG. et al. Reuniting public health and medicine: the university of new mexico school of medicine public health certificate. Am J Prev Med. 2011;41(4 Suppl 3):S214–S219.

10. Dodson S, Gleason AW. Web 2.0 support for residents' and fellows' patient care and educational needs. Med Ref Serv Q. 2011;30(2):95–101.

11. Pozdol JR. Ten steps to increase library impact on an academic health sciences campus.Med Ref Serv Q. 2010;29(3):229–239.

12. Kroth PJ, Phillips HE, Eldredge JD. Leveraging change to integrate library and informatics competencies into a new CTSC curriculum: a program evaluation. Med Ref Serv Q. 2009;28(3):221–234.

13. Jeffery KM, Maggio L, Blanchard M. Making generic tutorials content specific: recycling evidence-based practice (EBP) tutorials for two disciplines. Med Ref Serv Q.2009;28(1):1–9.

14. Eldredge JD. Student Peer Assessment as an Instructional Strategy. Albuquerque, NM: LOEX 37th National Conference; 2009.

15. Chen HC, Tan JP, O'Sullivan P, Boscardin C, Li A, Muller J. Impact of an information retrieval and management curriculum on medical student citations. Acad Med.2009;84(10 Suppl):S38–S41.

16. Eldredge JD, Carr R, Broudy D, Voorhees RE. The effect of training on question formulation among public health practitioners: results from a randomized controlled trial.J Med Libr Assoc. 2008;96(4):299–309.

17. Ryce A, Dodson S. A partnership in teaching evidence-based medicine to interns at the university of Washington medical center. J Med Libr Assoc. 2007;95(3):283–286.

18. Teal J, Eldredge J. Staff Development. Higginbottom PC. Chicago: Medical Library Association; 2005. The University of New Mexico.

19. Reavie K, Persily GL, Souza KH. In: A guide to developing end user education programs in medical libraries. Connor E, editor. New York: Haworth Information Press and Haworth Medical Press; 2005. Integrating Medical Informatics into the School of Medicine Curriculum at the University of California, San Francisco; pp. 209–225.

20. Owen DJ, Persily GL, Babbitt PC. In: A guide to developing end user education programs in medical libraries. Connor E, editor. New York: Haworth Information Press and Haworth Medical Press; 2005. An Informatics Course for First-Year Pharmacy Students at the University of California, San Francisco; pp. 129–143.

21. Eldredge JD. In: Informatics in health sciences curricula. Sewell RR, Brown JF, Hannigan GG, editor. Chicago: Medical Library Assn; 2005. EBM informatics component of the Genetics & Neoplasia Block.

22. Eldredge JD. Search strategies for population and social subjects in a medical school curriculum. Med Ref Serv Q. 2004;23(4):35–47.

23. Eldredge JD. The librarian as tutor/facilitator in a problem-based learning (PBL) curriculum. Ref Serv Rev. 2004;32(1):54–59.

24. Brown JF, Nelson JL. Integration of information literacy into a revised medical school curriculum. Med Ref Serv Q. 2003;22(3):63–74.

25. Kaplowitz J, Wilkerson L. Reaching and teaching new medical students. Acad Med.2002;77(11):1173.

26. Kaplowitz JR, Yamamoto DO. Web-based library instruction for a changing medical school curriculum. Libr Trends. 2001;50(1):47–57.

27. Eldredge JD, Rhyne RL. In: Handbook on problem-based learning. Rankin JA, editor. Chicago: Medical Library Assn; 1999. Building foundations for effective library skills in medical education: library/biometry projects in the first month of medical school; pp. 407–432.

28. Eldredge JD, Teal JB, Ducharme JC, Harris RM, Croghan L, Perea JA. The roles of library liaisons in a problem-based learning (PBL) medical school curriculum: a case study from university of New Mexico. Health Libr Rev. 1998;15(3):185–194.

29. Owen DJ. Using personal reprint management software to teach information management skills for the electronic library. Med Ref Serv Q. 1997;16(4):29–41.

30. Butros A. Using electronic mail to teach MELVYL MEDLINE. Med Ref Serv Q.1997;16:69–75.

31. Minchow RL, Pudlock K, Lucas B. Breaking new ground in curriculum integrated instruction. Med Ref Serv Q. 1993;12:1–18.

32. Eldredge J. A problem-based learning curriculum in transition: the emerging role of the library. Bull Med Libr Assoc. 1993;81(3):310–315.

33. Hannigan GG. Action research: methods that make sense. Med Ref Serv Q.1997;16(1):53–58.

34. Vezzosi M. In: Chandos information professional series. Connor E, editor. Oxford: Chandos Pub; 2007. Evidence-based librarian ship : case studies and active learning exercises; pp. 19–40.

35. Somekh B. In: The Sage handbook of educational action research. Noffke SE, Somekh B, editor. Thousand Oaks, CA: Sage Publications Ltd; 2009. Introduction; pp. 1–9.

36. Association of Academic Health Sciences Libraries. Annual Statistics of Medical School Libraries in the United States and Canada. 32. Seattle: AAHSL; 2010.

37. Eldredge JD, Heskett KM, Henner T, Tan J. New Horizons for Integrating Library/Informatics Skills into Medical Curricula: Report of the 2010 LiME Research Study. Stanford, CA: New Horizons: Selecting, Teaching and Inspiring the Next Generation of Physicians Association of American Medical Colleges Western Group on Educational Affairs (AAMC/WGEA) Regional Conference; 2011.

38. National Board of Medical Examiners. USMLE moves to next step in design, review.Examiner. 2008;55(2):1–4.

39. Anderson OS. Changing the USMLE: challenges and opportunities for physiology and other medical school basic science departments. Physiologist. 2009;52(2):39–44.

40. Association of Academic Health Sciences Libraries. AAHSL Membership Directory 2009 and AAHSL 30th Annual Report, 2007–2008. Seattle, WA: Association of Academic Health Sciences Libraries; 2009. pp. 5–29.

41. Eldredge JD, Henner T, Heskett K, Tan J. Current Practices in Library/Informatics Instruction in Academic Libraries Serving Medical Schools in the Western US. Asilomar, CA: Health and Interprofessional Education for the Underserved: Model Programs and Innovations Association of American Medical Colleges Western Group on Educational Affairs (AAMC/WGEA) Regional Conference; 2010.

42. Liaison Committee on Medical Education. Functions and Structure of a Medical School. Standards for Accreditation of Medical Education Programs Leading to the M.D. Degree.2010.

43. Gardois P, Colombi N, Grillo G, Villanacci MC. Implementation of Web 2.0 services in academic, medical and research libraries: a scoping review. Health Info Libr J.2012;29(2):90–109.

44. Lubans J. Educating the library user. New York: R. R. Bowker Co; 1974.

45. Smith HC. A course director's perspectives on problem-based learning curricula in biochemistry. Acad Med. 2002;77(12 Pt 1):1189–1198.

46. Burrows S, Ginn DS, Love N, Williams TL. A strategy for curriculum integration of information skills instruction. Bull Med Libr Assoc. 1989;77(3):245–251.

47. Allegri F. Course integrated instruction: metamorphosis for the twenty-first century. Med Ref Serv Q. 1986;4(4):47–66.

48. Traditi LK, Le Ber JM, Beattie M, Meadows SE. From both sides now: librarians' experiences at the rocky mountain evidence-based health care workshop. J Med Libr Assoc. 2004;92(1):72–77.

49. Dorsch JL, Jacobson S, Scherrer CS. Teaching EBM teachers. Med Ref Serv Q.2003;22(2):107–114.

50. Farber E. Faculty-librarian cooperation: a personal retrospective. Ref Serv Rev.1999;27(3):229–234.

51. Travis TA. Librarians as agents of change: working with curriculum committees using change agency theory. New Dir Teach Learn. 2008;2008(114):17–33.

52. Wiggins ME. Instructional design and student learning. Ref Serv Rev. 1999;27(3):225–228.

53. Krathwohl DR, Bloom BS, Masia BB. Taxonomy of educational objectives: the classification of educational goals. Handbook 2: Affective domain. New York: McKay; 1964.

54. Curzon SC. In: Integrating information literacy into the higher education curriculum: practical models for transformation. 1. Rockman IF, editor. San Francisco: Jossey-Bass; 2004. Developing faculty-librarian partnerships in information literacy; pp. 29–45.

55. Shurtz S. Thinking outside the classroom: providing student-centered informatics instruction to first- and second-year medical students. Med Ref Serv Q. 2009;28(3):275–281.

56. Moore M. Teaching physicians to make informed decisions in the face of uncertainty: librarians and informaticians on the health care team. Acad Med. 2011;86(11):1345.

57. Lodenius L, Honkanen M. Medical information specialist as teacher: teaching searching skills. J Eur Assoc Health Inf Libr. 2011;7(3):3–12.

6

CHAPTER

A CASE STUDY: PLANNING A STATEWIDE INFORMATION RESOURCE FOR HEALTH PROFESSIONALS: AN EVIDENCE-BASED APPROACH

Erinn E. Aspinall, MSI, AHIP; Katherine Chew, MLS; Linda Watson, MLS, AHIP, FMLA; Mary Parker, MA

Health Sciences Libraries, University of Minnesota Twin Cities, 450B Diehl Hall/505 Essex Street Southeast, Minneapolis, MN 55455 Cooperative Purchasing and Electronic Resources Services, Reference Services, Continuing Education, Minitex, University of Minnesota, 15 Andersen Library, 222 21st Avenue South, Minneapolis, MN 55455-0439

ABSTRACT

Question

What is the best approach for implementing a statewide electronic health library (eHL) to serve all health professionals in Minnesota?

Setting

The research took place at the University of Minnesota Health Sciences Libraries.

Methods

In January 2008, the authors began planning a statewide eHL for health professionals following the five-step process for evidence-based librarianship: formulating the question, finding the best evidence, appraising the evidence, assessing costs and benefits, and evaluating the effectiveness of resulting actions.

Main Results

Conclusion

1. STATEMENT OF THE CASE

Area-wide electronic health libraries (eHLs) that provide broad access to clinical-level information—such as electronic books, journals, guidelines, and drug information—are becoming more common on national and international levels. The growing number of eHLs reflects the increased awareness that health professionals require access to the latest evidence-based information in order to provide quality care. In response to this recognized need, the Health Sciences Libraries (HSL) at the University of Minnesota conducted an evidence-based feasibility study between January and September 2008 to determine the best approach for implementing an eHL that would serve all health professionals in Minnesota. The work was guided by the mission of the Minnesota eHL project, which was to foster clinical excellence by providing equitable access to quality, evidence-based health information for all of Minnesota's health practitioners, researchers, and students and to provide accurate health information to every citizen of Minnesota so they can become engaged in the patient-care partnership and make informed decisions.

2. SETTING

The eHL project was coordinated by the HSL in close partnership with Minitex. The HSL plays a prominent role in providing health information outreach to the state. It supports the goal of the University of Minnesota Academic Health Center to "expedite the dissemination and application of new knowledge into the promotion of health and delivery of health care in Minnesota" [1]. The HSL is a Resource Library for the National Library of Medicine's National Network of Libraries of Medicine, with an additional designation as an Outreach Library. The HSL also serves the state through its support of the Academic Health Center's land grant mission, which includes the primary role of educating health care professionals and generating and disseminating new knowledge to improve the health of Minnesotans

Minitex is a publicly supported network of academic, public, state government, kindergarten-through-twelfth-grade school, and special libraries working cooperatively to improve library services for their users in Minnesota, North Dakota, and South Dakota. It is an information- and resource-sharing program of the Office of Higher Education and the University of Minnesota Libraries that is funded by the Minnesota Legislature. The Minnesota State Library Services, a unit of the Minnesota Department of Education, provides additional funding to support services for Minnesota libraries.

With the combined goals of disseminating new knowledge to promote health and improving library services for Minnesotans, a partnership between the HSL and Minitex was a natural fit for developing an eHL for the state. The HSL's partnership with Minitex provided the added benefit of building on an established infrastructure made available through the Electronic Library for Minnesota (ELM) <http://www.elm4you.org>. ELM is an information portal that provides streamlined access to information resources for Minnesotans via statewide Internet protocol (IP) authentication.

The partnership between the HSL and Minitex provided sufficient capacity to support an eHL for the state. The timing for the project was also found to be appropriate in terms of infrastructure, need, stakeholder support, and strategic alignment. Regarding infrastructure, research has indicated that the state's health professionals and Minnesota households have sufficient access to computers and the Internet [3,4]. Regarding need, inequalities in patient care have been seen both nationally and throughout Minnesota [5–8]. Additionally, the state's 40,000 health professions students and 160,000 licensed health professionals have benefited from different levels of access to evidence-based clinical information, depending on their institutional affiliations [9,10]. Regarding stakeholder support, the concept of a statewide eHL has received strong support from university administration, health care organizations and representatives, and health sciences and other libraries across the state. Finally, regarding strategic alignment, the eHL project aligned with several health care reform initiatives in the state during 2007 and 2008 [5,11,12]. The final reports from these initiatives incorporated language that related to the importance of evidence-based health information, the involvement of patients in the health care process, and the goal of implementing a statewide electronic health record system that could serve as a delivery mechanism for eHL materials. This evidence was taken into account when determining the viability for an eHL in Minnesota as each of these aspects would impact its ultimate success.

3. METHODOLOGY

The eHL project activities followed the five-step evidence-based librarianship (EBL) process as defined by Eldredge [13]:

1. formulate a clearly defined, relevant, and answerable question;

2. search for an answer in both the published and unpublished literature, plus any other authoritative resources, for the best available evidence;

3. critically appraise the evidence;

4. assess the relative value of expected benefits and costs of any decided upon action plan; and

5. evaluate the effectiveness of the action plan.

3.1 Formulating the question

The question formulation process was guided by the EBL setting, perspective, intervention, comparison, and evaluation (SPICE) template for question building [14]. Using this framework, the following question was developed to guide the eHL planning process:

What is the best model for providing equitable access to relevant information resources for all health professionals in Minnesota, as compared to the best practices used by existing area-wide eHLs from outside of the state, that would align with local needs and resources?

In this case, the setting is Minnesota, the perspective is health professionals, the intervention is the model of equitable access, the comparison is existing best practices, and the evaluation is alignment with local needs and resources.

3.2 Finding the evidence

After developing a structured question to guide the EBL process, work was done to find relevant evidence. Because evidence was lacking in traditional publication venues, a multistep process was employed to locate other information sources that could address the original question. This included an environmental scan and an information resource assessment. These tasks were aimed at gathering best practices in the following areas: audience or population served, information resources, technology and access, funding model, and implementation and sustainability.

3.3 Performing an environmental scan

The environmental scan was a two-step process that included a competitive analysis and a questionnaire that surveyed health sciences librarians about projects that license clinical information for unaffiliated health professionals [15]. For the competitive analysis, a review of projects that provided area-wide services to non-affiliates was completed. A total of nine projects were identified through an Internet search using selected keywords with the Google search engine (Table 1). The project websites were reviewed to gather information related to population served and eligibility, available information resources, technology and access restrictions, and funding models. This information was placed in a matrix, and an additional column was added to capture information relating to the implementation and continued sustainability of eHLs. A literature review was also conducted as part of the competitive analysis to identify information that would supplement the findings from the review of project websites. A total of twelve articles were located by searching library literature databases (Library, Information Science & Technology Abstracts, Library Literature & Information Science) and medical literature databases (PubMed), as well as gray literature (Google Scholar), on keywords related to the selected projects (e.g., AZHIN, OhioLINK) [15]. New information located through this process was added to the competitive analysis matrix in the categories described above.

For the second part of the environmental scan, a questionnaire was developed to serve as an additional means of gathering information about statewide initiatives outside of Minnesota that license clinical information for unaffiliated health professionals [16]. The 36-question instrument focused on best practices in the following categories: population served, information resources, technology and access, funding models, and sustainability. An additional comments section was included to gather information not addressed in the structured questions. This questionnaire was reviewed and approved by the HSL director. Institutional review board approval for the survey protocol was granted through the University of Minnesota's Office of the Vice President for Research, and the questionnaire was distributed via a web-based survey tool. Respondents were recruited based on their membership in the Association of Academic Health Sciences Libraries (AAHSL). The AAHSL email discussion list was used to send out an announcement of the upcoming survey, the questionnaire, and 2 reminder notices. This procedure followed the Dillman total design survey method, modified to accommodate a quick turn-around time [17]. Forty-eight of the 143 AAHSL member institutions responded to the survey, for a response rate of 33%. Following the close of the survey, aggregate and individual data were generated using the web-based survey tool. The survey data were analyzed by the project manager. Key findings were summarized, and identifiable information was removed before it was distributed to the project team and the AAHSL email discussion list in a final report [16].

Table 1. Electronic health library (eHL) competitive analysis projects

Project name/organization	Uniform resource locator (URL)	Audience
Arizona Health Information Network (AZHIN)/University of Arizona Health Sciences Library	http://www.azhin.org	Serves the Arizona Health Sciences Library, the major teaching hospitals in Arizona, The University of Arizona College of Medicine, and the Arizona Area Health Education Centers
Electronic Health Library of BC (e-HLbc)	http://www.ehlbc.ca	Serves 6 British Columbia (BC) health authorities, 24 publicly funded postsecondary institutions 3 provincial ministries, the College of Physicians and Surgeons of BC, and the Physiotherapy Association of BC
Georgia Interactive Network (GaIN)/Mercer Medical Library	http://gainweb.mercer.edu	Serves health care institutions in Georgia, including over 50 institutional members, representing hospitals, clinics, and public health departments
HEAL-WA/University of Washington Health Sciences Library	http://www.heal-wa.org	Serves specified, licensed health care professionals in Washington state
Library Consortium of Health Institutions in Buffalo (LCHIB)	http://hubnet.buffalo.edu	Serves individuals affiliated with hospitals, health sciences schools, health sciences libraries, and other health-related organizations throughout western New York
OhioLINK	http://www.ohiolink.edu	Serves 16 public or research universities, 23 community or technical colleges, 50 private colleges, and the State Library of Ohio
Prepaid Articles Service at Medical College of Wisconsin (MCW) Libraries	http://www.mcw.edu/mcwlibraries/prepaidarticles.htm	Serves the MCW, Children's Hospital of Wisconsin, and Froedtert Hospital
TexShare	http://www.texshare.edu	Serves over 700 public and academic libraries and libraries of clinical medicine in Texas
Virtual Library of Virginia (VIVA)	http://www.vivalib.org	Serves Virginia's 39 state-assisted colleges and universities, 33 private, nonprofit institutions, and the Library of Virginia

3.4 Analyzing Information Resources

As part of the information resource analysis, a questionnaire was developed to identify a set of resources that would meet the information needs of the state's health professionals [18]. The methodology for the information resources survey matched that of the best practices survey described above, including the process for the survey design, approval, implementation, analysis, and distribution of results. The 19-question instrument focused on topics related to project support, current resource usage, information needs and gaps, and best practices, and it included a comments section to gather information not addressed in the structured questions. Respondents were recruited based on their membership in the Health Sciences Libraries of Minnesota (HSLM) association. Thirty-five of 71 HSLM members responded to the survey, for a response rate of 46%.

The information resources survey results were combined with the findings of a 2005 survey on the information needs of community-based preceptors working for the University of Minnesota Academic Health Center [3]. Questions in this survey related to the value of access to online resources, resources currently available, and ranking of specific resources. The nearly 500 respondents represented a broad range of health professionals, including family practitioners, nurses, physical and occupational therapists, genetic counselors, medical technicians, and pharmacists. Additionally, usage statistics were analyzed to identify the high-use resources that the HSL's patrons access.

The findings from the environmental scan and information resource analysis were used to identify resources, access, pricing, and technology requirements that addressed local needs. Emails were then sent to vendors to inquire about their ability to meet these criteria. Vendors had varied responses to the emails. While some expressed initial support, others voiced concerns about losing individual subscriptions, incurring liability for providing clinical information to health consumers, and controlling access. In some cases, initial emails were followed by conference calls with vendor representatives to provide additional details and to address concerns. After the initial and follow-up conversations with vendors, a request for information (RFI) was issued through the University of Minnesota's Purchasing Services to gather structured information on the capacity of information vendors to respond to the eHL's specific project requirements [19].

4. RESULTS

4.1 Best Practices

The multistep process used to gather evidence related to eHL implementation helped identify best practices in the areas of population served or audience, technology and access, funding model, and implementation and sustainability (Table 2). In summary, it was found that eHLs typically serve health professionals based on institutional affiliation, that the service is usually provided by academic health sciences libraries, that a mixed model approach is typically used to fund eHLs, and that eHLs should be supported by two to five full-time equivalents and employ a governing body for oversight.

Table 2. Best practices for eHLs

Category	Best practices
Audience or population served	▪ eHLs typically restrict access to a defined set of health professionals and often require that individuals belong to a member institution to benefit from services.
Information resources	▪ eHLs typically license resources that are evidence based and that support the information needs of a broad range of health care providers.
Technology and access	▪ eHLs are typically coordinated by academic health sciences libraries. ▪ Onsite access is typically granted to member organizations through Internet protocol (IP) verification. ▪ Offsite (remote) access is typically provided by member organizations to their affiliates through the use of a proxy server, when available.
Funding model	▪ eHLs are typically funded using a mixed-model approach, with revenue coming from membership fees, grants, and government funding. ▪ eHLs are typically more sustainable over time when they receive the bulk of their financial support from recurring state funds, though they often supplement costs in other ways (e.g., grants, membership fees).
Implementation and sustainability	▪ eHLs have typically received recurring state funds from departments of health and/or education. ▪ eHLs would ideally employ between 2 and 5 full-time equivalents. ▪ eHLs are typically guided by governing bodies, with work being carried out by the project team and by subcommittees working toward a specific charge. ▪ eHLs typically provide value-added services, either for free or at an added cost.

Through the RFI process, it was found that several information vendors were able to respond favorably to the project requirements that reflected the information needs of health professionals in the state, as well as specific technology, access, and cost requirements. The criteria included in the RFI stated that the suite of resources provided by the vendors must:

- represent the needs of a broad range of health professionals (physicians and nurses in particular);

- provide access to the following categories of resources: evidence-based medicine and evidence-based nursing point-of-care products, clinical drug references, full-text medical and nursing electronic journals, full-text medical and nursing electronic books, general medical and nursing bibliographical databases, and the Cochrane Library;

- be provided at a realistic and feasible cost, to be evaluated based on the annual price per health professional or health professions student user;

- be accessible via the ELM portal's statewide IP authentication system;

- allow unlimited access or access for a large number of concurrent users; and

- be reasonable for a staff of one to two full-time equivalents to manage (Note: one to two full-time equivalents were detailed in the RFI, as opposed to the two to five recommended by best practices because a portion of the workload would overlap with the current responsibilities of the HSL and Minitex staff) [19].

4.2 Appraisal of the evidence

Once the best practices were identified, work was done to appraise the evidence. The appraisal was conducted by comparing the best practices with the mission of the eHL, while taking into consideration the project team's expertise as health information professionals working in Minnesota. During the appraisal process, it was found that the timing for eHL implementation in Minnesota was appropriate in terms of capacity, infrastructure, need, stakeholder support, and strategic alignment, as described in the "Setting" section of this case study. However, the best practice of providing access based on institutional affiliation would significantly limit the project's mission of providing equitable access for all of Minnesota's health professionals. Additionally, institutional-based access would limit use by the state's health consumers. As a result, this best practice would not support the project's goal of providing accurate health information to the citizens of the state so that they can become engaged in the patient-care partnership and make informed health decisions.

The findings of the appraisal process led to the formation of recommendations that were based on the best practices gathered through the EBL process, modified to reflect the project goals and knowledge of the local environment (Table 3). Most notably, the eHL project team recommended that resources should be selected based on the needs of the project's primary audience, the state's licensed health professionals, but that access should be granted to all health practitioners, researchers, and students, regardless of institutional affiliation, and to all of Minnesota's five million citizens.

Table 3. Minnesota eHL recommendations

Category	Recommendations
Audience or population served	■ The eHL should be made available to all of Minnesota's health professionals (i.e., health practitioners, researchers, and students) regardless of their institutional affiliation.
	■ The eHL should be made available to all of Minnesota's 5 million citizens.
	■ The primary audience for the eHL should be all licensed health professionals in Minnesota.*
Information resources	■ eHL resources should be relevant to health professionals. Resources that are specifically aimed at a general (consumer) audience should not be considered.*
	■ eHL resources should represent the needs of a broad range of specialists (e.g., administrators, physicians, nurses, pharmacists, mental/behavioral health specialists), and physicians and nurses in particular.*
	■ eHL resources should support evidence-based practice.*
	■ Specific resources should be selected with input from stakeholders.
	■ The eHL should provide access to following categories of resources: evidence-based medicine/point-of-care product, evidence-based nursing/point-of-care product, clinical (professional-level) drug reference, full-text medical and nursing electronic journals, full-text medical and nursing electronic books, general medical and nursing bibliographical databases, and the Cochrane Library.
Technology and access	■ The University of Minnesota University Libraries, represented by the Health Sciences Libraries (HSL) and Minitex, should coordinate statewide access to resources.*
	■ The eHL should be made available through the Electronic Library for Minnesota (ELM), which is operated and administered by Minitex.
	■ eHL resources should support IP access. Resources requiring individual login/password should not be considered.*
	■ eHL resources should allow access for every citizen of Minnesota. Resources that limit use to health professionals or health professions students should not be considered.
	■ eHL resources should allow unlimited access or access for a large number of concurrent users.
Funding model	■ The funding model should cover the costs of the licensed resources, as well as the staff to support initial implementation and long-term maintenance.
	■ A mixed funding model with multiple revenue streams should be employed to support the eHL.*
	■ Recurring state funds should be sought as one eHL revenue stream.*
	■ Varying methods of recurring funds should be explored with stakeholders to see which model(s) might be best received by the legislature and other funding agencies.
	■ Selected information resources must be provided at a realistic/feasible cost, to be evaluated based on a $5.00–$10.00 annual price per health professional/health professions student.
	■ License agreements should be negotiated for a 3–5 year period to leverage buying power.
Implementation and sustainability	■ The funding model should include support for maintaining Go Local as part of the My Health Minnesota suite of resources.
	■ The eHL should be supported by a minimum staff of 2 full-time equivalents.*
	■ The eHL should be guided by a governing body made up of key stakeholders.*
	■ The work of the eHL should be carried out by subcommittees with representatives from medical and other libraries.*

* Based on best practices.

5. DISCUSSION

The EBL process of formulating the question, finding the best evidence, and appraising the evidence provided sufficient data to develop recommendations for eHL implementation in Minnesota. However, as indicated by EBL guidelines, it is important to examine the value of the data by assessing costs and benefits and evaluating the effectiveness of resulting actions.

5.1 Assessing costs and benefits

With recommendations in hand, work was done to evaluate their feasibility through a cost-benefit analysis. The eHL promised many positive results in terms of social benefits. These included contributing to an integrated and evidence-based health care system in Minnesota, which would enable patients to become partners in care, support continuous learning through the education process and into clinical practice, encourage recruitment and retention of Minnesota's rural health professionals, and support collaborative and practice-based health research, among others. The eHL would also have positive financial benefits as it would leverage economies of scale and buying power for equitable access to resources, leverage existing investments in information management and technology provided by the HSL and Minitex, and ultimately, reduce health care costs.

In addition to these social and financial benefits, it was determined that eHL implementation would carry little risk, as it was supported by best practices, reflected the needs of the state, and was fiscally responsible. However, the eHL recommendations were built on certain assumptions that would have to be addressed to ensure successful implementation. This was done by completing a risk assessment that detailed project assumptions along with their related mitigating actions and dependencies (Table 4).

5.2 Evaluating the results

The planning process for the eHL concluded in September 2008, with implementation set to begin in spring 2009. The timetable incorporated a 2-phase implementation schedule. Phase I of the plan was estimated to cost $1.2 million a year, based on the numbers provided by vendors in their responses to the RFI. With approximately 200,000 primary users of eHL resources (i.e., licensed health professionals and enrolled health professions students), this equated to an estimated annual cost per user of $6.00. These funds were to be requested for a 2-year period from Minnesota's health systems with additional contributions from the University of Minnesota Academic Health Center. This shared cost-funding model would have supported the licensing of both an evidence-based medicine and an evidence-based nursing product that would be made available to Minnesota's health professionals and to every citizen of the state.

The two years of funding provided by Minnesota's health systems would have allowed for an outcomes-based evaluation of the eHL, determining its successes and areas for improvement. The use of a proof-of-concept approach that would incorporate evaluation measures would illustrate the value of the eHL to the state's governing bodies and help prepare for phase II of the project, which would include a legislative proposal for recurring state funds in 2011.

Table 4. Assumptions related to eHL implementation

Assumption	Mitigating actions	Dependencies
The partnership between the HSL and Minitex will continue.	Development of a memorandum of understanding to be signed by both partners.	Partners can identify agreeable terms.
Resources will be made available through the ELM portal, operated and administered by Minitex.	Identification of vendors who can work within the eHL model.	Vendors are willing and able to provide access according to the eHL's "equal access" model.
Recurring funds for the eHL will be secured.	Development of a legislative proposal for eHL support.	Buy-in from key stakeholders in the academic health center, the University of Minnesota, and elsewhere.
The funding model will cover the costs of resources and staff.	Incorporation of salary for 2 full-time equivalents; negotiation of fixed pricing, secured through a formal request for proposal (RFP) process.	Resource pricing and salary estimates are accurate and will remain relatively stable over time.
Resources will be licensed for an annual cost of $5–$10 per health care user, which is equivalent to a total annual cost of $1–$2 million.	Selection of resources to match projected costs.	Vendor quotes align with cost estimates.
The eHL will work under a model of shared governance, made up of a steering committee and subcommittees.	Identification of the makeup of the steering committee and initial contact to test feasibility.	Support from health care organizations, health professions programs, and health sciences and other libraries across the state.
User support will be coordinated by the HSL but carried out by partner libraries in Minnesota.	Development of an informal partnership agreement that identifies an eHL champion in each partner organization.	Support from health sciences and other libraries across the state.

The timing for the two-phase implementation plan coincided with the severe economic downturn that occurred in fall 2008 that affected not only the university, but the state's health systems as well. As a result, the implementation process has been delayed until the state's economy proves to be more stable.

While the delayed implementation of the eHL limited formal evaluation of outcomes, the EBL process was deemed successful for several reasons. First, it was found that the question formulated by following EBL guidelines was indeed answerable. Second, by comparing the findings with professional expertise and the mission of the eHL project, recommendations were generated that aligned with local needs and resources, thereby satisfying the evaluation measure identified in the original EBL SPICE question. Finally, the recommendations formed the basis of a plan that was sensitive to possible risks, that has proved to remain relevant over time, and that can be put into practice when resources become available.

CONCLUSION

Through the EBL process, data were gathered that strengthened the authors' confidence in their ability to select, license, and deploy a suite of electronic resources that aligned with best practices and met local needs. The evidence-based feasibility study also showed the importance of integrating sustainability planning into an eHL project in order to support long-term success. This was done by engaging stakeholders in the planning process, ensuring adequate capacity and infrastructure to support an eHL, and aligning the proposal with statewide health initiatives. As resources become available to implement Minnesota's eHL, additional work will be done to evaluate the program and illustrate the extent to which equitable access to clinical information can support quality, cost-effective health care.

ACKNOWLEDGMENTS

The authors acknowledge the work done by the Electronic Health Library of British Columbia (e-HL*bc*). The e-HL*bc*'s Business Case and other supporting documents were valuable references throughout the feasibility study for an eHL in Minnesota.

REFERENCES

1. University of Minnesota, Academic Health Center. AHC strategic plan [Internet] The University; 2005. [rev 20 May 2008; cited 27 Apr 2009].

2. University of Minnesota, Academic Health Center. AHC strategic planning process phase II - report on defining question no. 1: what is our role in the health of Minnesotans; our land grant mission? [Internet] The University; 2000. [rev 7 Jun 2000; cited 27 Apr 2009].

3. University of Minnesota, Academic Health Center, Minnesota AHEC, Bio-Medical Library.Survey of preceptors for the health

sciences programs at the University of Minnesota The University; 2005.

4. Center for Rural Policy and Development. The 2007 Minnesota Internet survey: tracking the progress of broadband [Internet] The Center; 2008. [rev Jun 2008; cited 27 Apr 2009].

5. Health Care Transformation Task Force. Health Care Transformation Task Force recommendations submitted to: Governor Tim Pawlenty and the Minnesota State Legislature [Internet] The Task Force; 2008. [rev Jan 2008; cited 27 Apr 2009].

6. Mangione-Smith R, DeCristofaro A.M, Setodji C.M, Keesey J, Klein D.J, Adams J.L, Schuster M.A, McGlynn E.A. The quality of ambulatory care delivered to children in the United States. N Engl J Med. 2007 Oct 11;357(15):1515–23.

7. McGlynn E.A, Asch S.M, Adams J, Keesey J, Hicks J, DeCristofaro A, Kerr E.A. The quality of health care delivered to adults in the United States. N Engl J Med. 2003 Jun 26;348(26):2635–45. [PubMed]

8. Minnesota Community Measurement. Minnesota HealthScores [Internet]. [cited 27 Apr 2009].

9. Minnesota Office of Higher Education. Enrollment data search [Internet]. The Office [rev 2006; cited 27 Apr 2009].

10. State of Minnesota, Health Licensing Boards. Biennial reports July 1, 2004–June 30, 2006: table I: licensing and registration summary [Internet] The State; 2006. [rev 30 Jun 2006; cited 27 Apr 2009].

11. The Legislative Commission on Health Care Access. Final report: recommendations submitted to the Minnesota State Legislature [Internet] The Commission; 2008. [rev Feb 2008; cited 27 Apr 2009].

12. Minnesota Department of Health, Minnesota e-Health Initiative Advisory Committee. A prescription for meeting Minnesota's 2015 interoperable electronic health record mandate. a statewide implementation plan [Internet] The Department; 2008. [rev 2008; cited 27 Apr 2009].

13. Eldredge J.D. Evidence-based librarianship: an overview. Bull Med Libr Assoc. 2000 Oct;88(4):289–302.

14. Booth A. Formulating answerable questions. In: Andrew B, Anne B, editors. Evidence based practice for information professionals: a handbook. London, UK: Facet Publishing; 2004. pp. 61–70.

15. Aspinall E.E. 2008. My health Minnesota: electronic health library: environmental scan [Internet]. [rev 3 Feb 2008; cited 27 Apr 2009].

16. Aspinall E.E. 2008. My health Minnesota: electronic health library: best practices survey [Internet]. [rev 21 Apr 2008; cited 27 Apr 2009].

17. Dillman D.A. Mail and telephone surveys: the total design method. New York, NY: John Wiley & Sons; 1978.

18. Aspinall E.E. 2008. My health Minnesota: electronic health library: information needs assessment [Internet]. [rev 30 Apr 2008; cited 27 Apr 2009].

19. Aspinall E.E. 2008. My health Minnesota: electronic health library: request for information [Internet]. [rev 16 Jun 2008; cited 27 Apr 2009].

7

CHAPTER

EVIDENCE-BASED PRACTICE INSTRUCTION BY FACULTY MEMBERS AND LIBRARIANS IN NORTH AMERICAN OPTOMETRY AND OPHTHALMOLOGY PROGRAMS

Katherine A. MacDonald, BA, BSc, MLIS; Patricia K. Hrynchak, OD, FAAO; Marlee M. Spafford, OD, PhD, FAAO

Associate Dean of Science, Undergraduate Studies, and Professor, School of Optometry and Vision Science; University of Waterloo, 200 University Avenue West, Waterloo ON, N2L 3G1, Canada

ABSTRACT

North American optometry and ophthalmology faculty members and vision science librarians were surveyed online (14% response rate) about teaching evidence-based practice (EBP). Similar to studies of other health care programs, all five EBP steps (Ask, Acquire,

Appraise, Apply, Assess) were taught to varying degrees. Optometry and ophthalmology EBP educators may want to place further emphasis on (1) the Apply and Assess steps, (2) faculty- and student-generated questions and self-assessment in clinical settings, (3) online teaching strategies, (4) programmatic integration of EBP learning objectives, and (5) collaboration between faculty members and librarians.

1. INTRODUCTION

Evidence-based medicine was first developed for physicians in the early 1990s 1. Since then, other health care providers have integrated this concept into their disciplines, and the phrase has broadened to evidence-based practice (EBP) 2–6.

Optometrists and ophthalmologists provide eye and vision care. Optometry educators are increasingly recognizing the importance of EBP in optometric education. North American EBP competencies for optometrists can be found in accreditation standards and educational competency statements 7, 8. Curricula are changing to incorporate EBP knowledge and skill development at both the undergraduate and continuing education levels 2, 9–11. Articles in the optometric literature address EBP's nature, value, and role in the profession as well as barriers to and deficits in its current use in the profession 2. EBP literature in ophthalmology education predates that in optometric education 12–15. As a result, ophthalmology may be further ahead than optometry in adopting and integrating EBP into its educational programs.

EBP is normally taught by both faculty members and librarians, although their differing expertise likely informs what, who, and how they teach 16–23. Surveying ophthalmology and optometry programs can provide useful insights for vision science librarians and faculty members who are interested in introducing EBP training into or enhancing it in their programs. The results should allow comparison of vision-related EBP education in optometry and ophthalmology with that employed by other health care professions. This study is the first to survey both librarians and faculty members regarding educational practices with respect to EBP education.

2. METHODS

2.1 Measures

After obtaining institutional ethics clearance, online surveys (Appendix A and Appendix B, online only) were developed, tested, and administered in 2011 for two North American cohorts: (1) optometry and ophthalmology faculty members and (2) vision science librarians. The survey content was developed based on a literature review 16, 17, 19, 20, 24–29, a previous unpublished survey by the first author, the knowledge and experience of the research team members, and consultations with several vision science librarians. The surveys were pilot-tested using a group of local health sciences or medical librarians and optometry faculty members.

The surveys contained twenty-seven Likert-scale or open-text survey questions that addressed teaching EBP. Questions sought information on respondents' demographics, teaching or learning methods and assessment, and institutional characteristics. Likert-scale questions were either four points (e.g., frequency) or six points (e.g., agreement). Twenty-four of the survey questions were the same for the faculty members and librarians (although six questions contained slight wording differences to fit the unique roles and responsibilities of the respondent groups). A chi-squared test was used to examine differences between cohorts that might emerge because of their differing expertise. In many academic institutions, librarians are faculty members, but for the purpose of this study, the authors separated the two groups to gain a functional perspective.

2.2 Sample

Potential survey participants were identified through purposeful and snowball sampling using Internet searches, email queries, and word of mouth. Canadian and US optometry and ophthalmology faculty members involved in teaching EBP were identified by searching the websites of schools and colleges offering doctor of optometry programs and medical schools and facilities offering ophthalmology programs. Vision science librarians were identified through the Association of Vision Science Librarians (AVSL) and/or through websites of institutions offering optometry or medical programs. The websites of 20 optometry and 103 ophthalmology schools or programs were searched, and potential respondents were identified for each site.

Potential respondents were sent a preliminary email to ascertain if they taught EBP and if not, were requested to pass the email on to the appropriate person. Potential participants were asked to respond to the preliminary email. The email addresses of all initial potential participants who did not respond to the preliminary email plus the email addresses of suggested participants formed the final survey participant list. The email addresses of potential participants who indicated that they did not teach EBP or did not want to participate were taken off the list. This list was forwarded to the University of Waterloo Survey Research Centre, who implemented the online survey. Participants were told the study wanted to explore if and how optometry and ophthalmology programs taught their students the EBP process.

3. RESULTS

3.1 Survey response

Seven of the 460 distributed surveys were returned due to email address errors or extended out-of-office notices, leaving a potential sample of 453 people (328 faculty members and 125 librarians). Sixty-six surveys were returned with 4 being significantly incomplete (2 faculty members and 2 librarians), leaving 62 complete surveys (14% response rate). The respondents included 34 faculty members (11%) and 28 librarians (24%).

Fifty (81%) of the respondents indicated they taught at least some of the 5-step EBP process as defined by Straus and Prasad (Ask, Acquire, Appraise, Apply, Assess) 30, 31, and 12 (19%) indicated they did not teach any of the steps. This article reports on the 50 completed surveys from respondents who taught the EBP process (24 faculty members and 26 librarians). The survey was not designed to determine whether the respondents came from the same or differing institutions.

Faculty member respondents were primarily optometrists (75%) who provided clinical care (88%) and had more than 16 years of teaching experience and taught in optometry programs (79%). Most librarian respondents had a master of library and information science degree (92%), more than 13 years of teaching experience, and taught in more than 1 type of health care program (92%).

Respondents obtained EBP training via 1 or more methods, including through self-directed learning (86%), from courses outside a degree program (50%), in a graduate degree program (20%), or in their professional training (18%).

Generally, 80% of respondents worked in professional programs affiliated with a college or university. Sixty-six percent of the respondents described their libraries as multidisciplinary health sciences or discipline-specific.

3.2 Instruction of evidence-based practice (EPB) steps

All 5 steps of the EBP process were taught but to varying degrees among the respondent cohorts (Table 1), based on frequency-based responses. All librarian respondents taught Ask and Acquire, with decreasing proportions teaching Appraise, Apply, and Assess. Overall, Assess was taught by the smallest proportion of all respondents (42%). Faculty member respondents were more likely to teach the steps Apply (83% versus 35%) and Assess (63% versus 23%) than librarians ($\chi^2 \leq 8.0$, $P \leq 0.01$).

Table 1. Evidence-based practice (EBP) steps taught by respondents

EBP element taught	Total (n=50)		Faculty (n=24)		Librarian (n=26)	
	n	(%)	n	(%)	n	(%)
Ask (Convert need for information into an answerable question)	43	(86.0%)	17	(70.8%)	26	(100.0%)
Acquire (Find best evidence with which to answer question)	46	(92.0%)	20	(83.3%)	26	(100.0%)
Appraise (Critically appraise evidence for validity, impact, and applicability)	37	(74.0%)	21	(87.5%)	16	(61.5%)
Apply (Integrate evidence with clinical expertise and patient values)	29	(58.0%)	20	(83.3%)	9	(34.6%)
Assess (Evaluate own effectiveness)	21	(42.0%)	15	(62.5%)	6	(23.1%)

3.3 Learner type and duration taught

The librarian and faculty member respondents taught a variety of learners. Using frequency-based responses, in decreasing order, the top 4 types of learners receiving EBP instruction were other faculty members (52%), optometry students (50%), medical students (48%), and non-ophthalmology residents or fellows (46%). Twenty four percent of respondents taught ophthalmology residents. The statistically significant instructor-cohort differences among respondents were that optometry students were more likely to receive instruction from faculty members (88%) than librarians (15%) (χ^2=29.6, P<0.01), while medical students, non-ophthalmology residents or fellows, and graduate students were more likely to receive instruction from librarians (77%, 77%, 58%, respectively) than faculty members (17%, 13%, 21%, respectively) ($\chi^2 \leq 7.06$, $P \leq 0.01$). The number of hours of EBP instruction, using a yes/no response, was 10 hours or less (31%), 11–20 hours (27%), and more than 20 hours (43%).

3.4 Settings

The 50 respondents taught EBP in a variety of settings. Using frequency-based responses, the top 4 EBP teaching settings were: small classrooms (64%), seminar rooms (60%), offices (54%), and clinics (52%). The significant instructor-cohort setting differences were that faculty members (79%) were more likely than librarians (27%) to teach in clinics (χ^2=13.7, P=0.01), while librarians (77%) were more likely than faculty members (8%) to teach in computer labs (χ^2=26.4, P=0.01).

3.5 Teaching methods

Most often, respondents employed lectures (72%); individualized instruction (68%); print or electronic tutorials, handouts, and guides (64%); small group case-based learning (62%); and live demonstrations of tools, resources, and processes (62%), according to frequency-based responses. There were several significant instructor-cohort method differences. Relative to faculty members, librarians were more likely to employ tutorials, handouts, and guides (print or online information modules, exercise sheets, or collections of resources around a specific topic) (89% versus 38%); live demonstrations (92% versus 29%); and practical sessions (hands-on computer lab exercises) (77% versus 33%) ($\chi^2 \geq 0.62$, $P \leq 0.01$). Faculty members were significantly more likely than librarians to utilize case discussions (63% versus 23%) and to teach in the course of patient care (67% versus 12%) ($\chi^2 \geq 8.0$, $P \leq 0.01$).

3.6 Teaching tools and aids

Respondents used anytime (asynchronous) more frequently than live (synchronous) web-based tools with the most prevalent tools being course management software (54%), online tutorials and modules (44%), and email (30%). The top 4 teaching aids developed were EBP resource lists (50%), search strategy worksheets (42%), question development sheets (38%), and EBP subject guides (34%). Librarian respondents were more likely than faculty members to use search strategy worksheets (63% versus 21%) and subject specific guides (54% versus 13%) ($\chi^2 \geq 8.49$, $P \leq 0.01$).

3.7 Assessment strategies

To assess student learning, the 4 most commonly reported assessment strategies using frequency-based responses were final exam questions (54%), EBP worksheets (42%), critical appraisal exercises (42%), and case reports (40%). Significantly more faculty members than librarians reported using case study reports (58% versus 23%) and observations of the students' the clinical practices (63% versus 12%) ($\chi^2 \geq 6.5$, $P \leq 0.01$). Librarians (58%) were more likely than faculty members (17%) to use library research assignments involving self-assessment of literature searching skills ($\chi^2 = 9.0$, $P \leq 0.01$). Of interest is that only 26% of all respondents used a critically appraised topic (CAT) report as an assessment method. A CAT is a 1-page summary of a patient-stimulated EBP learning effort that includes the clinical question, the bottom line, an evidence summary, comments, and citations.

3.8 Collaboration and teaching support

Frequency-based questions were used to ascertain collaboration levels and activities. More respondents collaborated with faculty members (72%) than librarians (56%), with significantly more librarians (77%) than faculty members (33%) collaborating with librarians ($\chi^2 = 9.62$, $P = 0.01$). When asked about engaging with colleagues to teach EBP, significantly more librarians than faculty members collaborated by discussing teaching strategies and assessment outside class (73% versus 25%), teaching a section of the course alone (77% versus 21%), or co-teaching in the classroom (62% versus 8%), laboratory or clinic (23% versus 0) ($\chi^2 \geq 6.3$, $P \leq 0.01$). These options were not mutually exclusive. In response to a yes/no question, both faculty members and librarians used campus teaching support mechanisms, such as instructional technologies services (25%) and teaching support services (10%), in a limited way.

3.9 Integration of EBP into program

EBP training was embedded into the learning environment through a variety of strategies. According to frequency-based questions, the 4 most common avenues employed for all respondents were in courses (80%), via individual consults (80%), by continuing education courses (54%), and as program milestones or other form of graduation requirements (34%). None of the cohort differences were statistically significant.

The level of EBP integration into programmatic curricula varied. Indicators of EBP integration most often took the form of EBP-related learning objectives in courses (68%). Less often, EBP was a programmatic milestone requirement (40%); incorporated into the programmatic mission, goals, and objectives (40%); or part of clinic experience learning objectives (36%).

4. DISCUSSION

The findings of this study suggest that training for optometry students and ophthalmology residents should address all five steps of the EBP process. Ophthalmology residents may actually receive relatively more EBP training than optometry students, because studies indicate that EBP training normally starts during undergraduate medical school 16, 22, 23. Librarian respondents were most likely to report teaching the earlier EBP steps, while faculty members reported teaching across the five steps. This result might reflect the need for clinician-based activity to inform teaching and assessment especially in the Apply step. Librarian involvement in these types of activities would likely be limited. The final Assess step was taught the least frequently by both groups; its absence may reflect time constraints, a lack of teaching methods, or the assumption that it occurs automatically. Unfortunately, in the absence of this final step, students might not learn how to review and refine their EBP process, making them less efficient at and reflective about the process.

The respondents in this survey reported similar approaches to educators in other disciplines in terms of teaching settings 16, 19, 20, 24, 27 and the use of multiple teaching methods 19, 24,27. Lectures, with or without interactive methods 16, 19, 20, 24, 27, 32–34, were most commonly used; however, this method has been shown not to change practice behavior in postgraduate education 35. The majority of respondents also used individual consultation, although less frequently. This approach has been shown to be a good way to introduce, reinforce, and master EBP skills 24, 32, 36, 37. Results suggest that individual consultation needs to be introduced or expanded if the integration of EBP into clinical practice for optometry and ophthalmology is to be achieved.

Over half the respondents taught EBP in clinical settings, and this finding aligns with the growing recognition that EBP instruction should be clinically based, so that the questions coming out of clinical interactions are asked and answered 24. Answering patient-specific questions arising during clinical care has been shown to increase knowledge and change clinical decisions among residents 24 as well as medical students 38. Expanding teaching of EBP in the clinical setting should be encouraged. Only 27% of librarian respondents taught in clinics. Other studies note that the presence and participation of librarians in rounds and morning reports helps learners with developing questions, developing search strategies, and finding clinical evidence17, 39–41. Utilizing librarians in clinical settings is an opportunity that could be expanded by optometry and ophthalmology programs.

While the use of CAT reports is well documented in the literature as a method to help novices learn to ask a clinical question, review the literature, and summarize the best available research evidence on the subject 18, 30, 42–44, only just over one-quarter of respondents employed CATs as an instructional method. This assessment strategy was not one of the top four strategies used by respondents. More optometry and ophthalmology educators may want to explore the use of CATs, because patient-focused, self-directed, and personalized learning in the clinical setting should lead to a greater chance of having the EBP behaviors integrated into future practice 5, 24, 26, 36, 38, 42. Beyond the use of course management systems, other web-based teaching tools such as online tutorials or modules and videos could be explored by educators. These web-based tools 16, 31, 33, 42, 45–49 were used by few respondents but can be as effective as standard lectures for gaining knowledge and changing attitudes 50.

The most common assessment strategies used were final exam questions, EBP worksheets, critical appraisal exercises, and case reports. This is consistent with other disciplines 18–20, 27,42–44, 47, 51, 52. Assessment is often organized around a series of assignments and reports that focus on the various steps of EBP 20, 21, 25, 26. Active application of the process (the "shows how" level of Miller's pyramid) 53 has been assessed using clinical vignettes and standardized patients 27, 54–56 and is being used by some respondents. This is another avenue of exploration or expansion because of its value for assessing the Apply and Assess steps.

Integration of EBP into curriculum and programmatic goals or competencies occurred primarily at the level of the individual course rather than the program. While other health care professions talk about the integration of EBP into the curriculum, the degree of formal integration is unreported and requires further study.

Limitations of this study include the small number of respondents, particularly among faculty members and more specifically ophthalmology faculty members. Therefore, the findings might not reflect EBP instruction in North American optometry schools and ophthalmology programs. In institutions where librarians are faculty members, their status as faculty members rather than professional staff members might influence what EBP steps are taught and assessed. Another limitation is that our sampling technique might have incorrectly identified or missed individuals who were responsible for EBP education.

CONCLUSION

The findings of this survey-based study provide the first indication of EBP educational practices used by faculty members and librarians training optometry students and ophthalmology residents in North America. Optometry and ophthalmology educators may want to increase their emphasis on the Apply and Assess EBP steps to ensure that application and improvement of the EBP process are ingrained in students before they become practitioners. Some of these educators may want to enhance their use of: (1) faculty- and student-generated questions and self-assessment in the clinical setting, (2) online teaching strategies with assistance from teaching support services on campus, (3) greater integration of EBP learning objectives at a programmatic level, and (4) increasing collaboration.

ACKNOWLEDGMENT

Erin Harvey, Statistics and Actuarial Science, University of Waterloo, provided statistical analysis.

REFERENCES

1. Evidence-Based Medicine Working Group (November 1992) Evidence-based medicine: a new approach to teaching the practice of medicine. JAMA. 1992 Nov 4;268(17):2420–5.

2. Anderton PJ. Implementation of evidence-based practice in optometry. Clin Exp Optom.2007 Jul;90(4):238–43.

3. Hoge MA, Tondora J, Stuart GW. Training in evidence-based practice. Psychiatr Clin North Am. 2003 Dec;26(4):851–65.

4. Moch SD, Cronje RJ, Branson J. Part 1. undergraduate nursing evidence-based practice education: envisioning the role of students. J Prof Nurs. 2010 Jan;26(1):5–13.

5. Thomas A, Saroyan A, Dauphinee WD. Evidence-based practice: a review of theoretical assumptions and effectiveness of teaching and assessment interventions in health professions.Adv Health Sci Educ Theory Pract. 2011 May;16:253–76.

6. Kronenfeld M, Stephenson PL, Nail-Chiwetalu B, Tweed EM, Sauers EL, McLeod TCV, Guo R, Trahan H, Alpi KM, Hill B, Sherwill-Navarro P, Allen MP, Stephenson PL, Hartman LM, Burnham J, Fell D, Kronenfeld M, Pavlick R, MacNaughton EW, Nail-Chiwetalu B, Ratner NB. Review for librarians of evidence-based practice in nursing and the allied health professions in the United States. J Med Lib Assoc. 2007 Oct;95(4):394–407. DOI: http://dx.doi.org/10.3163/1536-5050.95.4.394.

7. Association of Schools and Colleges of Optometry. Attributes of students graduating from schools and colleges of optometry [Internet] The Association; 2011 [cited 5 Nov 2012].

8. Competence Committee of Canadian Examiners in Optometry. Competency-based performance standards for the Canadian standard assessment in optometry [Internet] The Association; 2005 [cited 15 Dec 2012].

9. Adams AJ. The role of research, evidence and education in optometry: a perspective. Clin Exp Optom. 2007 Jul;90(4):232–7.

10. Adams AJ. Whither goes evidence-based optometry. Optom Vis Sci. 2008 Jul;85(4):219–20.

11. Keller PR. The evidence in evidence-based practice. why the confusion. Clin Exp Optom.2012 Nov;95(6):618–20.

12. Burnier MN. The evaluation of scientific information in medicine. Can J Ophthalmol. 2004 Feb;39(1):5–6, 9.

13. Coleman AL. Applying evidence-based medicine in ophthalmic practice. Am J Ophthalmol.2002 Oct;134(4):599–601.

14. Lee AG, Boldt HC, Golnik KC, Arnold AC, Oetting TA, Beaver HA, Olson RJ, Carter K. Using the journal club to teach and assess competence in practice-based learning and improvement: a literature review and recommendation for implementation. Surv Ophthalmol.2005 Nov–Dec;50(6):542–8.

15. Wormald R. Bridging the gap to evidence-based eye care. Community Eye Health. 2004 Oct;17(51):40–1.

16. Lynn VA. Foundations of database searching: integrating evidence-based medicine into the medical curriculum. Med Ref Serv Q. 2010 Apr;29(2):121–31.

17. Hatala R, Keitz SA, Wilson MC, Guyatt G. Beyond journal clubs: moving toward an integrated evidence-based medicine curriculum. J Gen Intern Med. 2006 May;21(5):538–41.

18. Haines SJ, Nicholas JS. Teaching evidence-based medicine to surgical subspecialty residents. J Am Coll Surg. 2003 Aug;197(2):285–9.

19. Finkel ML, Brown HA, Gerber LM, Supino PG. Teaching evidence-based medicine to medical students. Med Teach. 2003 Mar;25(2):202–4.

20. Burns HK, Foley SM. Building a foundation for an evidence-based approach to practice: teaching basic concepts to undergraduate freshman students. J Prof Nurs. 2005 Nov–Dec;21(6):351–7.

21. Shlonsky A, Stern SB. Reflections on the teaching of evidence-based practice. Res Soc Work Pract. 2007 Sep;17(5):603–11.

22. MacEachern M, Townsend W, Young K, Rana G. Librarian integration in a four-year medical school curriculum: a timeline. Med Ref Serv Q. 2012;31(1):105–14.

23. Maggio LA, Tannery NH, Chen HC, ten Cate O, O'Brien B. Evidence-based medicine training in undergraduate medical education: a review and critique of the literature published 2006–2011. Acad Med. 2013 Jul;88(7):1022–8.

24. Schilling LM, Steiner JF, Lundahl K, Anderson RJ. Residents' patient-specific clinical questions: opportunities for evidence-based learning. Acad Med. 2005 Jan;80(1):51–6.

25. Brancato VC. An innovative clinical practicum to teach evidence-based practice. Nurse Educ. 2006 Sep–Oct;31(5):195–9.

26. Cayley WE., Jr Evidence-based medicine medical students: introducing EBM in a primary care rotation. Wis Med J. 2005 Apr;104(3):34–7.

27. Holloway R, Nesbit K, Bordley D, Noyes K. Teaching and evaluating first and second year medical students' practice of evidence-based medicine. Med Educ. 2004 Aug;38(8):868–78.

28. Jack BA, Roberts KA, Wilson RW. Developing the skills to implement evidence based practice—a joint initiative between education and clinical practice. Nurse Educ Pract. 2003 Jun;3(2):112–8.

29. Fineout-Overholt E, Johnston L. Teaching EBP: implementation of evidence: moving from evidence to action. Worldviews Evid Based Nurs. 2006;3(4):194–200.

30. Straus SE, Glaziou P, Richardson SW, Haynes RB. Evidence based medicine: how to practice and teach it. 4th ed. Toronto, ON, Canada: Churchill Livingstone; 2011.

31. Prasad K. Fundamentals of evidence-based medicine: basic concepts in easy language. New Delhi, India: Meeta Publishers; 2007.

32. Bookstaver PB, Rudisill CN, Rebecca Bickley A, McAbee C, Miller AD, Piro CC, Schulz R. An evidence-based medicine elective course to improve student performance in advanced pharmacy practice experiences. Am J Pharm Educ. 2011 Feb;75(1):9.

33. Pitkala KH, Mantyranta T, Strandberg TE, Makela M, Vanhanen H, Varonen H. Evidence-based medicine—how to teach critical scientific thinking to medical undergraduate students.Med Teach. 2000 Jan;22(1):22–6.

34. Potomkova J, Mihal V, Zapletalova J, Subova D. Integration of evidence-based practice in bedside teaching paediatrics supported by e-learning. Biomed Pap Med Fac Univ Palacky Olomouc Czech Repub. 2010 Mar;154(1):83–7.

35. Coomarasamy A, Khan K. What is the evidence that postgraduate teaching in evidence based medicine changes anything? a systematic review. BMJ. 2004 Oct 30;329(7473):1017–9.

36. Bradt P, Moyer V. How to teach evidence-based medicine. Clin Perinatol. 2003 Jun;30(2):419–33.

37. Elnicki DM, Kolarik R, Bardella I. Third-year medical students' perceptions of effective teaching behaviors in a multidisciplinary ambulatory clerkship. Acad Med. 2003 Aug;78(8):815–9.

38. McGinn T, Seltz M, Korenstein D. A method for real-time, evidence-based general medical attending rounds. Acad Med. 2002 Nov;77(11):1150–2.

39. Duggar DC, Christopher KA, FitzGerald L, Wood RT. Does providing onsite immediate access to electronic resources using a laptop and wireless network improve resident participation in morning report. J Hosp Lib. 2008 Nov;8(4):411–7.

40. Skhal KJ. A full revolution: offering 360 degree library services to clinical clerkship students. Med Ref Serv Q. 2008 Fall;27(3):249–59.

41. Weaver D. Enhancing resident morning report with "daily learning packages." Med Ref Serv Q. 2011;30(4):402–10.

42. Aronoff SC, Evans B, Fleece D, Lyons P, Kaplan L, Rojas R. Integrating evidence based medicine into undergraduate medical education: combining online instruction with clinical clerkships. Teach Learn Med. 2010 Jul;22(3):219–23.

43. Burneo JG. Jenkins ME. Teaching evidence-based clinical practice to neurology and neurosurgery residents. Clin Neurol Neurosurg. 2007 Jun;109(5):418–21.

44. Rugh JD, Hendricson WD, Glass BJ, Hatch JP, Deahl ST, 2nd, Guest G, Ongkiko R, Gureckis K, Jones AA, Rose WF, Gakunga P, Stark D, Steffensen B. Teaching evidence-based practice at the University of Texas Health Science Center at San Antonio dental school. Tex Dent J. 2011 Feb;128(2):187–90.

45. Jeffery KM, Maggio L, Blanchard M. Making generic tutorials content specific: recycling evidence-based practice (EBP) tutorials for two disciplines. Med Ref Serv Q. 2009 Spring;28(1):1–9.

46. Bradley P, Oterholt C, Herrin J, Nordheim L, Bjorndal A. Comparison of directed and self-directed learning in evidence-based medicine: a randomised controlled trial. Med Educ. 2005 Oct;39(10):1027–35.

47. Cook DA, Dupras DM. Teaching on the web: automated online instruction and assessment of residents in an acute care clinic. Med Teach. 2004;26(7):599–603.

48. Kaneshiro KN, Emmett TW, London SK, Ralston RK, Richwine MW, Skopelja EN, Brahmi FA, Whipple E. Use of an audience

response system in an evidence-based mini-curriculum.Med Ref Serv Q. 2008 Fall;27(3):284–301.

49. Genes N, Parekh S. Bringing journal club to the bedside in the form of a critical appraisal blog. J Emerg Med. 2010 Oct;39(4):504–5.

50. Davis J, Crabb S, Rogers E, Zamora J, Khan K. Computer-based teaching is as good as face to face lecture-based teaching of evidence based medicine: a randomized controlled trial. Med Teach. 2008;30(3):302–7.

51. Mottonen M, Tapanainen P, Nuutinen M, Rantala H, Vainionpaa L, Uhari M. Teaching evidence-based medicine using literature for problem solving. Med Teach. 2001 Jan;23(1):90–1.

52. Temple C. L. F, Ross DC. Acquisition of evidence-based surgery skills in plastic surgery residency training. J Surg Educ. 2011 May–Jun;68(3):167–71.

53. Miller G. The assessment of clinical skills competence performance. Acad Med. 1990 Sep;65(9):S63–S67.

54. Rutten GMJ, Harting J, Rutten STJ, Bekkering GE, Kremers SPJ. Measuring physiotherapists' guideline adherence by means of clinical vignettes: a validation study. J Eval Clin Pract. 2006 Oct;12(5):491–500.

55. Shah R, Edgar D, Evans BJW. Measuring clinical practice. Ophthalmic Physiol Opt. 2007 Mar;27(2):113–25.

56. Waxman KT. The development of evidence-based clinical simulation scenarios: guidelines for nurse educators. J Nurs Educ. 2010 Jan;49(1)):29–35.

8

CHAPTER

MEASURING PATRONS' TECHNOLOGY HABITS: AN EVIDENCE-BASED APPROACH TO TAILORING LIBRARY SERVICES

Jin Wu, MSIS; Amy J. Chatfield, MLS; Annie M. Hughes, MSLS; Lynn Kysh, MLIS; Megan Rosenbloom, MLIS

Norris Medical Library, University of Southern California, 2003 Zonal Avenue, Los Angeles, CA 90089

Information Services Librarian, Jennifer Ann Wilson Dental Library & Learning Center, Herman Ostrow School of Dentistry, University of Southern California, Los Angeles, CA 90089

Norris Medical Library, University of Southern California, 2003 Zonal Ave, Los Angeles, CA 90089

ABSTRACT

Librarians continually integrate new technologies into library services for health sciences students. Recently published data are lacking about student ownership of technological devices, awareness of new technologies, and interest in using devices and technologies to interact with the library. A survey was implemented at seven health sciences libraries to help answer these questions. Results show that librarian assumptions about awareness of technologies are not supported, and student interest in using new technologies to interact with the library varies widely. Collecting this evidence provides useful information for successfully integrating technologies into library services.

1. INTRODUCTION

Health sciences students enter programs with a variety of devices (laptops, tablets, and smartphones) as well as varying levels of knowledge of software. Academic medical libraries that serve students invest in new devices, software, and technology, and devote staff time to projects involving these tools 1–5. The University of Southern California (USC) Health Sciences Libraries (HSL) have conducted similar pilot projects using mobile devices, adding quick response (QR) codes to link print and electronic books; providing instruction on software, e-readers, and mobile devices; and establishing presences on social media networks. Not all of these projects succeeded as planned, so librarians turned to the literature for guidance.

Library literature provides data about patron ownership of devices and operating systems 6–10and patron interest in new software and technologies 11–16. The rates of adopting technologies and assessing the value of these new technologies in a library setting are also discussed 1–5,17–24. No articles simultaneously address what seem to be the three critical facets for designing student-focused library technology projects: device ownership, awareness of software and technology, and willingness to use devices and software to interact with the library. Moreover, given the rapid pace of technological change, the information in published studies is almost always too dated to be completely useful. Health sciences librarians require current data to use in making evidence-based decisions when considering projects using new technologies.

2. METHODS

In early 2012, the Emerging Technologies Committee of USC HSL developed a sixteen-question instrument (Appendix, online only). The survey was submitted to the Institutional Review Board (IRB) at USC, and the board concluded that the research was exempt from IRB review. The survey was generated using Wufoo <http://www. wufoo.com> and was distributed only in electronic form to the USC Health Sciences incoming students in fall 2012 via email discussion lists or during class orientations and library registration. Students were advised that the survey was entirely anonymous and optional, and intended to provide USC HSL with information to tailor library services to current student interests.

The committee developed a letter to introduce the idea of the survey to other health sciences libraries to achieve a more representative sample and to lay the groundwork for a common annual survey. The letter requested that no questions except those regarding status and affiliation with the school be altered, that the health sciences schools should utilize their own survey tools, and if they did not have one available, USC would host a survey for them and that the survey should be distributed close to the arrival of incoming students. The letter was sent via email to the 116 member libraries in the Association of Academic Health Sciences Libraries. Six health sciences libraries chose to participate. Each library designed its own method of distributing the survey to health sciences students.

3. RESULTS

Data were collected and combined from 1,513 respondents (out of 6,270 potential respondents) representing 7 institutions. Some institutions altered questions or responses rendering them unfit for inclusion in analysis. Response rates are included for each item. Other questions permitted selection of multiple answers, and total number of responses received is included. Some institutions expanded the survey beyond incoming health sciences students; all responses are analyzed here.

Respondents were professional students (69%), graduate students (15%), post-baccalaureate students (3%), faculty (4%), staff (1%), residents (5%), and other (1%, n=1,326). Respondents were asked to report their affiliations: medicine (25%), pharmacy (23%), dentistry (15%), nursing (15%), physical therapy (5%), public health (5%), science and health/biomedical (5%), veterinary medicine (3%), osteopathic medicine (2%), occupational therapy (1%), and other (1%, n=1,510). For ages, 55% were in the 20–25 range; 25% were 26–30 years, 7% were 31–35 years, 6% were 36–45 years, 5% were 46–

55 years, 2% were 56–65, and 1% were 66 and up (n = 1,350). There were more female respondents (59%) than male (41%, n = 1,513).

3.1 Ownership of devices

Responses indicated that incoming students used a wide variety of devices. Fifty-six percent of respondents used PC laptops, 46% Mac laptops, 74% smartphones, 34% tablet devices, and 15% e-book readers; none did not use any of these devices (n = 3,475). For smartphone operating systems, 48% had iOS, 25% had Google/Android, 6% had Blackberry OS, 2% had Windows Mobile, 1% had other, 4% were unsure, none had Symbian, 7% did not have a smartphone but planned to get one, and 6% were not interested in any smartphone (n = 1,521). When asked about tablet operating systems, 34% had iOS, 10% had Google/Android, 1% had Blackberry, 1% had Windows Mobile, 2% were not sure, 17% did not have a tablet but planned to get one, and 32% were not interested in tablets (n = 1,467).

Respondents were asked which brand of e-book reader they used. Twenty percent owned Kindles; 18% used their tablets for reading e-books; 3% owned Nooks; none owned Kobos, Sonys, Bookeens, or COOL-ERs; 12% did not have a reader but planned to get one, and 46% said that they were not interested in any e-book reader (n = 1,300).

3.2 Awareness of new technologies

Respondents were asked to select a phrase that best describes them with regard to use of technologies. Seven percent used new technologies before anyone else, 19% used technologies a little before others, 47% used technologies at the same time as everyone else, 25% took a while to use new technologies, and 2% avoided new technologies (n = 1,148).

When asked about preferences for e-books, 33% of respondents said that they did not use e-books at all and preferred print books. Twenty percent said that they intended to use e-books more in the future; 19% said that they preferred e-books for leisure reading, but not for academic purposes; 9% said they preferred print books for leisure reading, but not for academic purposes; and 5% said they intended to use more print books in the future. Eleven percent of respondents did not have a preference with regard to e-books (1,738 total responses). When asked to select all the instant messenger (IM) tools that they currently used, 62% used Facebook, 31% used Google Talk, 7% used Yahoo Messenger, 7% used other, none used Meebo or Pidgin, and 20% did not use any IM tools at all (2,148 responses).

To gauge awareness, respondents were asked to report if they used or had heard of Facebook, Twitter, Google+, MySpace, LinkedIn, Second Life, Delicious/Diigo, Zotero, Skype, FourSquare, Pinterest, Mendeley, Google Reader, and QR codes. The only one of these systems used regularly by more than half of respondents was Facebook: 59% report that they used it all the time, 24% used it sometimes, 5% were using it more, 6% used to use it, another 6% never used it, and no one had never heard of it (n = 1,301). There was some use of Twitter and Google+: 5% of respondents used Twitter all the time, 10% used it sometimes, 3% were using it more, 9% used to use it, 2% never heard of it, and 71% never used it (n = 1,278). For Google+, 11% used it all the time, 13% used it sometimes, 7% were using it more, 13% used to use it, 53% never used it, and 2% never heard of it (n=1,255). Ten percent of respondents used Skype all the time, while 36% used it sometimes, 11% were using it more, 20% used to use it, 22% never used it, and 1% never heard of it (n=1,277). Other options were rarely used: 67% never used MySpace, and 3% never heard of it (n=1,267); 62% never used LinkedIn, and 8% never heard of it (n=1,264); 47% never used Second Life, and 51% never heard of it (n=1,260); and 58% never used Pinterest, and 15% never heard of it (n=1,261). In several cases, more than half of the respondents had never heard of the option: 60% of respondents said they never heard of Delicious/Diigo, and 38% never used it (n=1,258). Thirty-five percent of respondents said they never used Zotero, and 60% never heard of it (n=1,346); and 39% never used Mendeley, and 58% never heard of it (n=1,261). FourSquare had 1% of respondents using it all the time, 2% using it sometimes, 1% using it more, 49% using it in the past, 24% never using it, and 24% never having heard of it (n=1,261). For Google Reader, 4% used it all the time, 10% used it sometimes, 5% were using it more, 5% used to use it, 52% never used, and 24% never heard of it (n=1,249). One percent used QR codes all the time, 6% used them sometimes, 3% were using them more, 4% used to use them, 39% never used them, and 47% never heard of QR codes (n= 1258).

3.3 Willingness to use new technologies to interact with the library

Respondents were asked how likely they would be to use their smartphone or tablet to interact with library services and staff (Figure 1). Respondents were most willing to use their smartphones or tablets to use medical apps provided through the library, check library hours, and use the library's electronic resources. They also were likely or fairly likely to use their smartphones or tablets to look for materials in the catalog or read e-books. When social media were examined, the majority of respondents were not interested in following the library on Twitter (31% unlikely, 46% extremely unlikely) or friending the library on Facebook (31% unlikely, 21% extremely unlikely).

Respondents were asked about their likelihood of using text/short message service (SMS) to receive library services. Signing up to receive overdue or renewal notices generated the most positive responses (30% extremely likely, 29% likely, 19% fairly likely, 9% unlikely, and 3% do not text [n=1,452]). There was moderate interest in texting a question to a librarian (13% would be extremely likely, 19% likely, 23% fairly likely, 18% unlikely, 18% extremely unlikely, and 3% do not text [n= 1,465]) or in texting to renew library material (32% would be extremely likely, 29% likely, 17% fairly likely, 9% unlikely, 10% extremely unlikely, and 3% do not text [n=1,452]). Little interest was evidenced in texting a call number from the catalog (16% would be extremely likely, 22% likely or fairly likely, 28% unlikely, and 22% extremely unlikely [n=1,444]).

The survey also asked respondents to indicate which monthly lunchtime workshops on technology topics they would be likely to attend. Google tools (39%) and presentation tools (37%) generated the most interest, followed by mobile device apps (29%), photo editing tools (28%), video editing tools (19%), social networking tools (15%), blogs (8%), really simple syndication (RSS) readers (7%), and Web 2.0 applications (6%) (n=3,010).

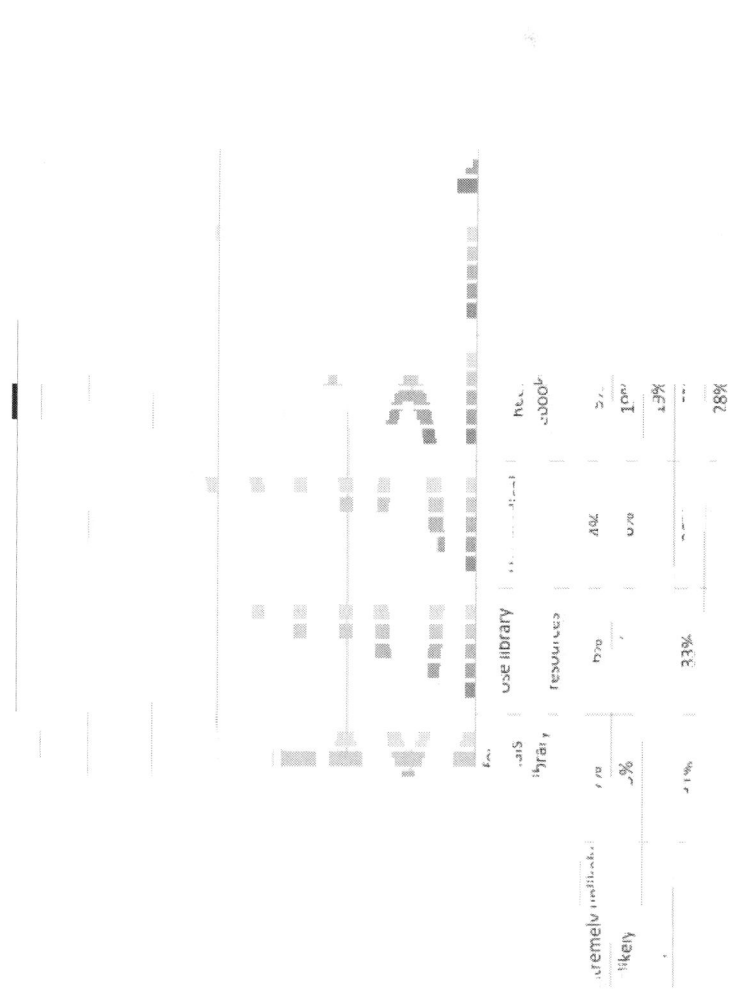

Figure 1. Likelihood of accessing library services through mobile technology

4. DISCUSSION

Despite the volume of library literature discussing the adoption and benefits of new technologies 1–5, 17–24, survey results of first-year students found that they do own devices but are unaware of many of the newer technologies. Regardless of their awareness of technologies, they are only willing to use certain types to interact with the library. There is clearly little interest in using social media with the library, but there is significant interest in using devices to communicate with the library to obtain information about hours, availability of materials, due dates and interest in accessing materials licensed by the library, such as e-books and apps. Health sciences students also perceive themselves to be equal to or slightly more advanced than others in terms of adopting new technologies.

While a general understanding of student technology awareness is helpful, it is the specifics of this kind of survey that make it so valuable. As intended, the survey data have been extremely helpful in planning new technology projects. For example, QR codes are perceived to be low-cost technologies to implement, but an 18-month project to create, print, and affix codes, and track usage for 200 print books and 10 subject guides took nearly 30 hours of staff time and the majority of the QR codes received no hits. Prior to the survey, librarians were unsure if low usage was due to placement and marketing or to lack of awareness of QR codes. Survey data made it clear that it was the latter, and so projects using QR codes—linking print books with electronic equivalents and affixing them in the stacks to guide browsers to subject-specific electronic books—have been eliminated.

The authors have also been able to use awareness data to select topics for drop-in educational workshops at the USC HSL. Sessions focused on technologies that patrons reported that they had never heard of, including e-books and LinkedIn, have been well attended. Additionally, survey results permitted an in-depth assessment of social media strategies. Efforts placed on using Google+, Facebook, and Twitter have been slowed and redirected to other communications mechanisms, given that respondents are unlikely to friend or follow the library on social networks. Staff time is being directed toward implementing the technologies that patrons are interested in, such as finding, downloading, and using medical apps for mobile devices.

The authors plan to administer this survey annually and hope that other institutions will do likewise. In that way, data from this survey can also be used to anticipate and monitor future interests and needs. E-book readers have been forecast as important devices, and USC HSL were considering purchasing several such devices. With a demonstrated lack of interest in such readers in 2012/13, this project can be placed on hold and revived if patron interest is shown to change. An annual survey will permit librarians to monitor technology adoption trends and react to these trends appropriately.

There are several limitations to the survey. The survey only examined first-year students. This population was chosen because the library could make changes that would affect this group; it can take time to launch new projects. Likewise, there are many surveys conducted on campus, and survey fatigue is a problem. From prior experience, the authors realized that asking all students to complete a survey would result in few responses from students who had progressed further in their degree programs. Another concern is that the majority of respondents were from medicine or pharmacy; however, when the authors examined enrollment, pharmacy and medicine students clearly compose at least 50% of the population that each library involved in the survey serves, so the response patterns match the population surveyed. The survey instrument did not include images. Several of the technologies, such as QR codes, may not be recognized by name but may be recognized visually. Images will be included in future surveys.

5. CONCLUSION

The survey reported here examines three facets of student use of technology to provide a better picture of patrons' technological habits: ownership of devices, awareness of new technologies, and willingness to use these technologies to interact with the library. The survey provides timely information that librarians can use to customize their services. It can easily be updated each year as new technologies emerge and older ones fade away without compromising the longitudinal data. Ample opportunities exist for comparing trends year over year in a single institution, as well as exploring trends via data sharing across institutions. Administering the survey annually can provide a portrait of how professional health sciences students differ from other populations in terms of their technological savvy and attitudes. With some alteration of questions, the survey can be extended to faculty and clinical user populations. Through these methods, librarians can base their services on real and timely evidence that comes straight from their patrons.

The authors of this survey will continue to administer it at USC HSL and are actively seeking additional collaborators. We will contact the email discussion list of the Association of Academic Health Sciences Libraries annually and discuss the survey at regional and national conferences to obtain additional participating institutions. Health sciences librarians are encouraged to contact the authors of this article to join the collaboration. The survey is updated each spring with new technologies, and the new instrument is distributed to institutional contacts during the late summer.

ACKNOWLEDGMENTS

The authors thank Janis F. Brown, AHIP, for her guidance with survey implementation and writing, and Emily Brennan, former librarian at the USC HSL, for her work developing and conducting the survey.

REFERENCES

1. Vucovich LA, Gordon VS, Mitchell N, Ennis LA. Is the time and effort worth it? one library's evaluation of using social networking tools for outreach. Med Ref Serv Q. 2013 Jan;32(1):12–25.

2. Garcia-Milian R, Norton HF, Tennant MR. The presence of academic health sciences libraries on Facebook: the relationship between content and library popularity. Med Ref Serv Q.2012 May;31(2):171–87.

3. Summey K, Richmond C, Bushhousen E. An examination of e-reader devices and their implications within an academic health science library. J Elec Res Med Lib. 2011 Jul;8(3):217–24.

4. Cuddy C, Graham J, Morton-Owens EG. Implementing Twitter in a health sciences library.Med Ref Serv Q. 2010 Oct; 29(4):320–30.

5. Hendrix D, Chiarella D, Hasman L, Murphy S, Zafron ML. Use of Facebook in academic health sciences libraries. J Med Lib Assoc. 2009 Jan; 97(1):44–7.

6. Le Ber JM, Lombardo NT, Honisett A, Jones PS, Weber A. Assessing user preferences for e-readers and tablets. Med Ref Serv Q. 2013 Jan; 32(1):1–11.

7. Khan N, Coppola W, Rayne T, Epstein O. Medical student access to multimedia devices: most have it, some don't and what's next. Inform Health Soc Care. 2009 Mar; 34(2):100–5.

8. Kennedy G, Gray K, Tse J. "Net generation" medical students: technological experiences of pre-clinical and clinical students. Med Teach. 2008 Feb; 30(1):10–6.

9. Dørup J. Experience and attitudes towards information technology among first-year medical students in Denmark: longitudinal questionnaire survey. J Med Internet Res. 2004 Jan–Mar; 6(1):e10.

10. Seago BL, Schlesinger JB, Hampton CL. Using a decade of data on medical student computer literacy for strategic planning. J Med Lib Assoc. 2002 Apr; 90(2):202–9.

11. Atlas MC. Are medical school students ready for e-readers. Med Ref Serv Q. 2013 Jan; 32(1):42–51.

12. Burgess K. Schick L. Assessment of medical students' use of e-books (originally submitted and listed in abstract book with the title "First-year medical student e-book survey") Poster presented at: One Health: Information in an Interdependent World; 113th Medical Library Association Annual Meeting and Exhibition, 11th International Congress on Medical Librarianship, 7th International Conference of Animal Health Information Specialists, and 6th International Clinical Librarian Conference; Boston, MA; May 3–8, 2013.

13. Bushhousen E, Norton HF, Butson LC, Auten B, Jesano R, David D, Tennant MR. Smartphone use at a university health science center. Med Ref Serv Q. 2013 Jan;32(1):52–72.

14. Franko OI, Tirrell TF. Smartphone app use among medical providers in ACGME training programs. J Med Sys. 2012 Oct;36(5):3135–9.

15. Von Muhlen M, Ohno-Machado L. Reviewing social media use by clinicians. J Am Med Inform Assoc. 2012 Sep–Oct;19(5):777–81.

16. Harris SS, Barden B, Walker HK, Reznek MA. Assessment of student learning behaviors to guide the integration of technology in curriculum reform. Info Serv Use. 2009 Jan;29(1):45–52.

17. Kraft M. Using quick response (QR) codes to discover e-books. Poster presented at: One Health: Information in an Interdependent World; 113th Medical Library Association Annual Meeting and Exhibition, 11th International Congress on Medical Librarianship, 7th International Conference of Animal Health Information Specialists, and 6th International Clinical Librarian Conference; Boston, MA; May 3–8, 2013.

18. Sevetson E, Boucek B. Keeping current with mobile technology trends. J Elec Res Med Lib.2013 Jan;10(1):45–51.

19. Chu SKW, Du HS. Social networking tools for academic libraries. J Lib Info Sci. 2012 Feb;45(1):64–75.

20. Sorensen K, Glassman NR. Point and shoot: extending your reach with QR codes. J Elec Res Med Lib. 2011 Jul;8(3):286–93.

21. Pacheco J, Kuhn I, Grant V. Librarians' use of Web 2.0 in UK medical schools: outcomes of a national survey. New Rev Acad Lib. 2010 Mar;16(1):75–86.

22. DeFebbo DM, Mihlrad L, Strong MA. Microblogging for medical libraries and librarians. J Elec Res Med Lib. 2009 Sep;6(3):211–23.

23. Lapidus M, Bond I. Virtual reference: chat with us. Med Ref Serv Q. 2009 May;28(2):133–42.

24. Kipnis D, Kaplan G. Analysis and lessons learned instituting an instant messaging reference service at an academic health sciences library: the first year. Med Ref Serv Q. 2008 Jan;27(1):33–51.

9

CHAPTER

DIGITAL EVIDENCE FOR DATABASE TAMPER DETECTION

Shweta Tripathi[1], Bandu Baburao Meshram[2]

[1]*Department of Computer Engineering, Fr. Agnel Institute of Technology, Navi Mumbai, India*
[2]*Head Department of Computer Technology, Veermata Jijabai Technological Institute, Mumbai, India*

ABSTRACT

Most secure database is the one you know the most. Tamper detection compares the past and present status of the system and produces digital evidence for forensic analysis. Our focus is on different methods or identification of different locations in an oracle database for collecting the digital evidence for database tamper detection. Starting with the basics of oracle architecture, continuing with the basic steps of forensic analysis the paper elaborates the extraction of suspicious locations in oracle. As a forensic examiner, collecting digital evidence in a database is a key factor. Planned and a modelled way of examination will lead to a valid detection. Based on the literature survey conducted on different aspects of collecting digital evidence for database tamper detection, the paper proposes a block diagram which may guide a database forensic examiner to obtain the evidences.

KEYWORDS

Tamper Detection; Log Files; Forensics; Oracle Database

1. INTRODUCTION

Database security is not a new buzz. But identifying the reasons of security violation in a database is a recent point of discussion. To understand how, when and what was tampered in a database the thorough knowledge of the architecture of database is required. To bind the vast information about architectures of different databases the architecture of oracle 10g is considered.

Database Forensics is a branch of digital forensic science relating to the forensic study of databases and their related metadata. For the forensic examination of a database, it has to be related to the timestamps that apply to the update time of a row in a relational table being inspected and tested for validity in order to verify the actions of a database user. Alternatively, a forensic examination may focus on identifying transactions within a database system or application that indicate evidence of wrong doing, such as fraud. Hence forth identifying who, when and how modified or tampered the data.

There are many approaches towards forensics of database defined by different researchers. A structured model to collect the evidence is a need of database forensics which has been proposed in this paper as the initial step of examining the database after the tamper.

The organisation of rest of the paper is as follows: Section 2 gives the literature survey. Based on basic concepts, section 3 is devoted for process to collect evidence. Section 4 gives the conclusion.

2. LITERATURE SURVEY

The literature survey explores from the basics of architecture of oracle 10g to the vulnerabilities in oracle, digital forensics analysis and database tamper detection.

2.1. Basics of Oracle Architecture

In this section the basics of oracle 10g is explained highlighting few **Figure 1** [1]. Oracle physical storage and memory structure is explained in brief. Since these locations may guide us to locate and analyse the tamper in the database.

2.1.1. Oracle Physical Storage Structures

The Oracle database uses a number of physical storage structures on disk to hold and manage the data from user transactions. Some of these storage structures, such as the datafiles, redo log files, and archived redo log files, hold actual user data; other structures, such as control files, maintain the state of the database objects, and text-based alert and trace files contain logging information for both routine events and error conditions in the database. **Figure 1** [1] shows the relationship between these physical structures and the logical storage structures.

2.1.2. Datafiles

One Oracle datafile corresponds to one physical operating system file on disk [1].

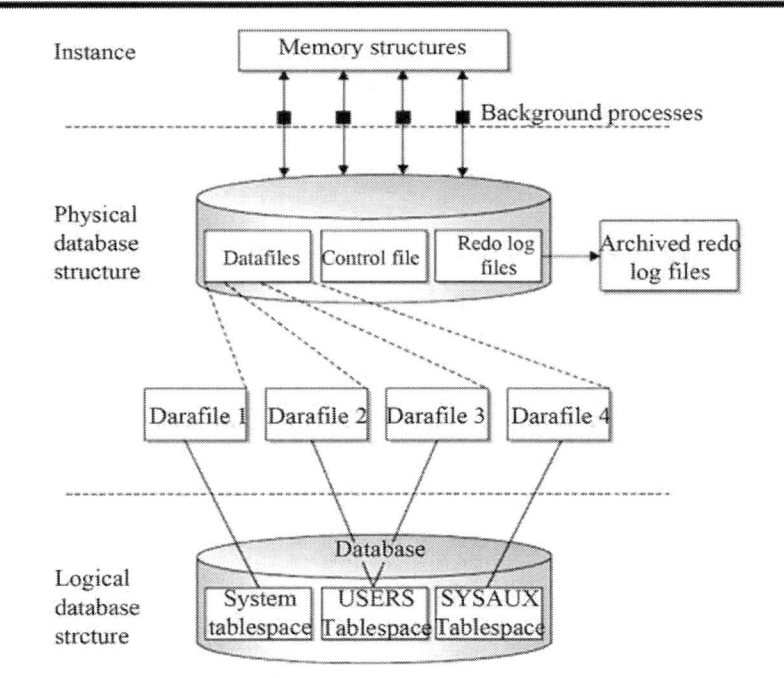

Figure 1. Oracle physical storage structures.

2.1.3. Redo Log Files

Whenever data is added, removed, or changed in a table, index, or other Oracle object, an entry is written to the current redo log file [1].

2.1.4. Control Files

Every Oracle database has at least one control file that maintains the metadata of the database (in other words, data about the physical structure of the database itself) [1].

2.1.5. Archived Log Files

An Oracle database can operate in one of two modes: archivelog or noarchivelog mode. When the database is in noarchivelog mode, the circular reuse of the redo log files is no longer available in case of a failure to a disk drive. Archivelog mode sends a filled redo log file to one or more specified destinations and can be available to reconstruct the database at any given point in time in the event that a database media failure occurs [1].

2.1.6. Alert and Trace Log Files

When things go wrong, Oracle can and often does write messages to the alert log and, in the case of background processes or user sessions, trace log files [1].

2.1.7. Backup Files

Backup files can originate from a number of sources, such as operating system copy commands or Oracle Recovery Manager (RMAN) [1].

2.2. Vulnerabilities in Oracle

To collect the evidences for data tamper detection one should have a brief knowledge of the suspicious location in the database. This part of the paper summarizes the vulnerabilities in oracle 10g with the [2].

The author mentions there are three areas an attacker can exploit to break in to a given system. Firstly, they can exploit the trust already given to them in the case of an inside attacker or in a social engineering attack; secondly, an attacker can exploit a weakness in the configuration of the server and third and last, an attacker can exploit a vulnerability in the software. Social engineering attacks are beyond the scope of this paper. We have highlighted few weaknesses in server configurations and software vulnerabilities. Some of the common configuration weaknesses that can be exploited to break in are listed below.

2.2.1. Exploiting Configuration Vulnerabilities

Exploiting configuration vulnerabilities is an easy task of an attacker due to following vulnerable configuration [2].

Default Usernames and Passwords Prior to 10g, Oracle was renowned for the number of default user accounts that have default passwords after an install. Some common accounts with their password include [2].

Username	Password
SYS	CHANGE_ON_INSTALL
SYSTEM	MANAGER

DBSNMP   DBSNMP

MDSYS   MDSYS

CTXSYS   CTXSYS

WKSYS   WKSYS

SCOTT   TIGER

Files: After installing 10g, both release 1 and 2, there are a number of files that contain the password selected by the DBA during installation. Whilst the passwords are obfuscated it is trivial to recover the clear text password from these files. If an attacker can gain access to these files then they have the keys to the kingdom. These files include:

CONFIGTOOLS.XML
CONFIGTOOLSINSTALLDATE.LOG
IAS.PROPERTIES
TARGETS.XML [2]

Reflections on default passwords Many times we often hear a DBA protest that they block direct connections to the database with a firewall rule setting—this generally means blocking access to port 1521—so they don't need to worry about default user IDs and passwords. However, they often forget to block ports 2100 and (less so) 8080 which means that, if the Oracle server is running the XML Database (XDB), then attacker can log in over these ports. An FTP service in XDB listens on TCP port 2100 and a web server on port 8080 [2].

Unnecessary Services Here is a simple chain of logic: the more services that are running, the greater the size of the attack surface; the greater the attack surface; the greater the likelihood of a flaw existing, the greater the likelihood of a successful break in. If there is no strict business requirement for something to be running, if it doesn't serve an active role in any business processes then it should be turned off [2].

The TNS Listener Prior to Oracle 10g the TNS Listener was installed without a password set for it. This means that, unless a password was set at a later time, anybody with access to port 1521, or whatever the port the server is listening on, could remotely administer the Listener. One of ways in which an attacker could exploit this was to set the listener's log file to a set location such as ~/.rhosts on a *nix system or a batch file in the Startup folder of the administrator on a Windows system. Once the location is set then the attacker can write content to the file such as "+ +" in the case of .rhosts or a command to execute in the case of a batch file in the Startup folder [2].

2.2.2. Exploiting Software Vulnerabilities

There are many different classes of software vulnerability and every so often new classes are discovered. Oracle has suffered from most at some point or continues to and these provide an attacker with a way into the server [2].

Buffer Overflow Vulnerabilities

A buffer overflow vulnerability is conceptually easy to understand. The programmer sets aside some memory to hold some data and they make some assumptions about the size of that data. Along comes an attacker however and stuffs too much data into the buffer. The buffer overflows and the data overwrites other "stuff" in memory—some of which may be crucial to the running of the program or the program's flow of execution. With this now under the control of the attacker they can get the program to do their bidding as opposed to what the program was supposed to do [2].

Format String Vulnerabilities

Format string vulnerability arises when a programmer uses one of the functions in the printf family of functions without specifying the format string. This, in effect, allows the attacker to present the format string instead [2].

PL/SQL Injection

PL/SQL injection vulnerabilities are one of the more common types of Oracle security flaw. Oracle stored procedures are known as packages. These packages contain data, procedures and functions. They also have the ability to execute dynamic SQL statements. By default, when a PL/SQL package executes it does so with the privileges of the owner and so, if the package suffers from a security vulnerability, it presents the attacker with the ability to run SQL as the owner of the package [2].

Trigger Abuse

Not only are packages, procedures and functions vulnerable to PL/SQL injection but triggers can be too. A trigger is a piece of PL/SQL code that fires when a specific event occurs—for example a user performs a DML operation such as an INSERT. A number of triggers have been found to contain security weaknesses. These include the SDO_LRS_TRIG_INS, SDO_GEOM_TRIG_INS1 and SDO_DROP_USER_BEFORE triggers owned by MDSYS and the CDC_DROP_CTABLE_BEFORE trigger [2].

Attacks via Oracle Application Server

Attacks from the outside of the network, particularly where the database server is properly protected by a firewall, typically originate from the web server—more often than not in the form of SQL injection attacks in the custom JSP, PHP or ASP application [2].

2.3. Digital Forensic Analysis Methodology

The entire investigation process can be divided into four phases [3].

1. Identification: in this phase it collects the information of the compromised system. System Configuration, software loaded, User profiles etc.

2. Collection Phase: collects the evidence from the compromised system. Evidence is most commonly found in files and Databases that are stored on hard drives and storage devices and media. If file deleted, recovering data from the deleted files and also collects evidence file deleted files.

3. Analysis phase: analyse the collecting data/files and finding out the actual evidence.

4. Report phase: the audience will be able to understand the evidence data which has been acquired from the evidence collection and analysis phases. The report generation phase records the evidence data found out by each analysis component. Additionally, it records the time and provides hash values of the collected evidence for the chain-of-custody.

2.4. Database Tampering

Maintaining data integrity is an important aspect of security in a database. The basics of database tampering can be explained better with the case study.

2.4.1. Example of Data Tampering

Any business cannot afford the risk of an unauthorized user observing or changing the data in their databases. There are several types of concerns that are realized about database security. They are "unauthorized data observation, incorrect data modification, and data unavailability". An example of unauthorized data observation would be a database user accessing information that they are not authorized to view. Incorrect data modification can be intentional or unintentional. Intentional data modification could be a student changing their grade or a data entry clerk accidentally entering the price of a line item incorrectly for an order. Data unavailability exists when "information crucial for the proper functioning of the organization is not readily available when needed." [4].

A data breach in a bank information system [5] was an unauthorized access to clients banking accounts which has resulted with money problems to a certain number of clients. Somehow, they have a minus on credit card accounts. Bank personnel are unable to identify the culprit. Persecution department send a team of digital forensics. They have a task to collect all evidence about suspicious transactions, examine them and make a report to the court of low. The team needs to follow a precisely defined procedure to provide valid court evidence (**Figure 2**).

Different kinds of log files are available in oracle 10g. To identify who had got the unauthorized access, when he had got the access and how he had tampered the data in the accounts database, a forensic examiner should collect the relative log files and perform the analysis to gain the information about the criminal.

2.4.2. Detection Methods

Different methods to detect tampering can be based on the identification of violation of the integrity of the database. The threats to integrity are [6] authentication and access control, application integration, database extensions, inherited OS vulnerabilities, privileged parties.

A novel relational hash tree [6] can be designed for efficient database processing, and evaluate the performance penalty for integrity guarantees. Such implementation does not require modification of system internals and can therefore easily be applied to any conventional DBMS.

Database provenance chronicles the history of updates and modifications to data [7]. Without additional protecttions, provenance records are vulnerable to accidental corruption, and even malicious forgery, a problem that is most pronounced in the loosely-coupled multi-user environments often found in scientific research. The problem of providing integrity and tamper-detection for database provenance can be solved by a checksum-based approach, is well-suited to the unique characteristics of database provenance, including non-linear provenance objects and provenance associated with multiple fine granularities of data [7].

The Enterprise Resource Planning system (ERP system) manages database through application on the database top layer [8]. The investigation of the corporate frauds, gives output as investigation data from application software, which cannot be trusted. The database design in the form of dual-book may be used for hiding data; hence investigation data cannot be obtained from these systems. In such case the data is not only provided in the application software in the database, but there is additional data which is related with investigation. Therefore, the investtigators must verify what investigation data exists on the database. These fields are based on the database and application. The attribute information of these fields are stored in a database schema, through the DBRE (database reverse engineering), it is possible to determine the table relationship by extracting schema information. The DBRE largely divide two parts: data structure extraction and data structure conceptualization. Improving the DBRE process and hence analyzing the table relationship of database and data extraction method is useful from the view of the digital forensics.

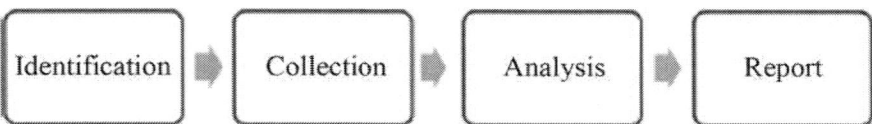

Figure 2. Investigation processes.

There are different mechanisms within a database management system (DBMS), based on cryptographically strong one-way hash functions, which prevent an intruder, including an auditor or an employee or even an unknown bug within the DBMS itself, from silently corrupting the audit log [9]. DBMS transparently store the audit log as a transaction-time database, so that it is available to the application if needed. The DBMS should also store a small amount of addition information in the database to enable a separate audit log validator (to be referred to simply as the validator) to examine the database along with this extra information and state conclusively whether the audit log has been compromised. The DBMS periodically send a short document (a hash value) to an off-site digital notarization service, to bind when changes were made to a database [10].

The above methods have discussed about the models to identify the threat and hence detect the evidences in the vulnerable database. But our problem as a database forensic examiner to analyse the data to identify who, where and how tampered the data, is still not solved. With above methods we cannot gather the data for analysis. The later part of the paper summarizers the work of the researchers in the field of identifying locations for collecting the evidences.

2.4.3. Sources for Evidences

The author of [11] has listed the main source of evidence as follows:

1 Listener log—logs connections to the listener, use lsnrctl to administrate it. It Can be found in /u01/app/oracle/oracle/product/10.2.0/db_4/network/listener.log.

2 Alert log—system alerts important to DB e.g. processes starting and stopping. It Can be found in /u01/app/oracle/admin/orcl/bdump 3) Sqlnet.log—some failed connection attempts such as "Fatal NI connect error 12170".

3 Redo logs—current changes that have not been checkpointed into the datafiles (.dbf).

   /u01/app/oracle/oradata/orcl/redo02.log

   /u01/app/oracle/oradata/orcl/redo01.log

   /u01/app/oracle/oradata/orcl/redo03.log 5) Archived redo logs—previous redo logs that can be applied to bring back the data in the db to a previous state using SCN as the main sequential identifier. This can be mapped to timestamp.

4 Fine-Grained Auditing audit logs viewable from FGA_LOG$ and DBA_FGA_AUDIT_TRAIL VIEW.

5 Oracle database audit SYS.AUD$ table and DBA_AUDIT_TRAIL VIEW.

6   Oracle   mandatory   and   OS   audit   /u01/app/oracle/
    admin/orcl/adump 9) Home-made trigger audit trails—bespoke
    to the system.

7   Agntsrvc.log—contains logs about the Oracle Intelligent agent.

8   IDS, Web server and firewall logs should also be integrated to the
    incident handling timeline. This will rely heavily on well
    synchronised time in the network as previously mentioned.

Different locations [12] to find forensics data is listed below:

- TNS listener log

- Many types of trace files

- Sqlnet logs (server and clients)

- Sysdba audit logs

- Datafiles for deleted data

- Redo (and archive) logs

- SGA (v$sql etc)

- Apache access logs

- v$db_object_cache

- Wrh$%% views

- Wri$ views

- Statspack views

- col_usage$

- Audit trails –

  - AUD$, FGA_LOG$

  - Application audit (who/when, triggers, other)

- Flashback, recycle bin The capabilities of LogMiners can be evaluated as a Forensics investigation tool. Its general applicability can be assessed for its forensics by testing how well it can create a timeline and copy of past database actions. LogMiner proves itself to be very useful for this purpose. The interpretation of the TIMESTAMP data types is also done by The LogMiners. Databases are excellent reporting mechanisms and LogMiner allows the analyst to use a database SQL interface to the redo logs of Oracle which are separate from the database itself.

At minimum this can provide verification of information found through normal database Auditing and at maximum could be the main source of information in an investigation and allow subsequent recovery of lost data [13].

3. PROPOSED SYSTEM

Our proposed system is based on two basic steps in identifying the location, that is summarizes the flow of processes for collecting the evidences and we have also designed the block diagram summarizing the vulnerable locations in the database.

3.1. Identifying Locations for Evidence

A series of papers on performing a forensic analysis of a compromised Oracle database server is published by Mr. Litchfield. Based on these papers we have identified the locations where we may obtain the evidences for tamper detection.

Redo Logs

A Redo Entry, otherwise known as a Redo Record, contains all changes for a given SCN [14]. The entry has a header and one or more "change vectors". There may be one or more change vectors for a given event. For example, if a user performs an INSERT on a table that has an index then several change vectors are created. There will be a redo and undo vector for the INSERT and then an insert leaf row for the index and a commit. Each change vector has its own operation code that can be used to differentiate between change vectors. The table below lists some of the more common ones:

5.1 Undo Record 5.4 Commit 11.2 INSERT on single row 11.3 DELETE 11.5 UPDATE single row 11.11 INSERT multiple rows 11.19 UPDATE multiple rows 10.2 INSERT LEAF ROW 10.4 DELETE LEAF ROW 13.1 Allocate space [e.g. after CREATE TABLE]

24.1 DDL The forensic examiner must go through each redo entry and work out what has happened and attempt to separate those which are "normal" and those which are part of an attack.

Data Blocks

The second paper [15] in this series is dedicated on Locating dropped objects. When the block is filled up, the server starts filling a new block. Each row in the block has a three byte header. The first byte is a marker and contains a set of flags to indicate the row's state. For example, if the row has been deleted the 5th bit of the byte is set. For example, a common set of flags value for a marker is $0 \times 2C$—which becomes $0 \times 3C$ when the "deleted" flag is set. This is an important point to remember as it is a key indicator when looking for dropped objects. The second byte of the row header is used to determine lock status and the third byte indicates the total amount of data in the row. If the total amount is greater than 255 bytes then the row header is four bytes allowing for up to 65,536 bytes. After the row header is the data itself. Each column of the row data is preceded with a byte indicating the size. If there is no data for a given column, in other words it is null, then it is represented with a $0 \times FF$.

TNS Listener's log file and the audit trail

To be able to log into the RDBMS an attacker [16] needs to know the Service Identifier or SID for the database. Before Oracle 10g this could be extracted from the TNS Listener with the SERVICES or STATUS command.

Here's something to be careful of with the audit trail. When a user successfully logs on a row is created in the audit trail. This has an ACTION# number of 100 (LOGON) and the TIMESTAMP# column reflects when the logon occurred.

In building a timeline of events this is important. This effectively hides when the user actually logged on. However, if we describe the AUD$ table we can see a LOGOFF$TIME column. If we then query this column, too, we can reconcile the logon and logoff times:

Live Response

When the database is shutdown cleanly this would wipe the audit trail making the task of the forensic examiner that little bit harder [17]. Of course, the attacker could do more than just wiped the audit trail in such a trigger. Due to issues like this and the loss of volatile information, some organizations prefer to perform an analysis on the system whilst it's still powered on and connected to the network. This is called a Live Response. Live Response is all about recovering and safely storing volatile data for later analysis, in other words, all the information that will disappear when the machine is disconnected from the network and switched off. Further, Live Response gives the forensic examiner the chance to collect non-volatile evidence in a "humanreadable" format that's easier to peruse than its stored binary version—for example event logs.

Views

There are a number of virtual tables and views that Oracle maintains for performance purposes [18]. These views are accessible to DBAs and can often contain evidence of attacks. Two of these views are of particular interest—V$SQL and V$DB_OBJECT_CACHE. The V$SQL fixed view contains a list of recently executed SQL. Evidence of an attacker's activities may be found in this fixed view and careful examination of the SQL_TEXT should reveal this. It must be stressed that if an attacker can find a way to execute arbitrary SQL as DBA, of which there are many, then they can clear the SQL from this view by executing "ALTER SYSTEM FLUSH SHARED_POOL". V$DB_OBJECT_CACHE contains details about objects in the library cache.

Oracle Recycle Bin

Whenever a table is dropped, the table and any [19] dependent objects such as indexes and triggers are moved to the Recycle Bin. This way, if it is decided that the table has been dropped in error, it can be recovered from the Recycle Bin using the UNDROP statement.

System Change Number

During a forensic examination of a compromised [20] Oracle database server the SCN and its timestamp can help tell the investigator whether a block of data has been changed. This is especially useful in those cases where there is an absence of other evidence such as the redo logs or audit trail. As with all forensic examinations it's critical not to change any evidence so any investigation should take place on a cold data file and not a live data file.

3.2. Steps to Collect the Evidence

With the help of the above study we have identified the steps which are useful in collecting the evidences [17].

1) Setup the evidence collection server by the following ways:

- firstly by mapping a drive if the system is running on Windows or has Samba and then using file redirection: D:\>listdlls.exe > z:\case-0001-listdlls.txt Using file redirection can be prone to error—for example the incident responder could type C instead of Z—which would be disastrous.

- The second method is to pipe output over the network using netcat or cryptcat.

2) Perform the following general steps to get basic information like [17].

System time and date:

The incident responder should first record the system time and date of system that they're investigating.

Logged on users:

The list of users that are currently logged on to the system and from where and for how long is extremely useful.

List all users and groups:

Obtain a list of all users, gathering details on when they last logged in, and groups on the server and group membership.

List open ports and connections :

All open and connected TCP ports should be collected as well as listening UDP ports.

List running processes:

A list of all running processes should be obtained. Close attention should be paid to suspicious looking entries and also any shells such as cmd.exe or /bin/sh—indeed keep an eye out for //bin/sh (note two slashes) as this may indicate an overflow or format string exploit has been launched. The forensic examiner should also get a list of each process's parent process.

List of DLLs or shared objects:

A list of the DLLs or shared objects that are loaded by each process should be obtained. Keep an eye out for odd looking names; on Windows look out for DLLs that are loaded via a UNC path across the network.

List of open handles:

As well as what file handles a process has open a list of other handles should be obtained as well. Whilst this can reveal what an attacker may have been doing it can also help identify "parentless" processes.

Perform memory dumps:

Memory dumps of all running process should be gathered even in what appear to be "normal" looking processes. The reason for this is to catch cloaking attacks—an attacker may launch a benign process like "notepad" and using CreateRemoteThread() load code into its address space.

Perform system memory dump

A dump of all system memory should be performed. This will cover those bit of memory not dumped when dumping each process.

Get file names and MACTimes:

The incident responder should perform a full recursive directory list of every disk and get file and directory names as well as their creation, access and modification times. They should also gather information about each file's owner and any special attributes such as whether the read only, system or hidden attributes are set.

Dump registry information:

On Windows all registry information should be dumped.

Locate and take copies of log files and message logs:

All of the servers log files and event and message logs should be copied to the collection server for analysis. These logs will vary from system to system depending upon what services are running.

3) Collect the Oracle files of Interest The Oracle specific log, trace and control files can be located in various places [17]. Firstly we need to know where each instance of Oracle is installed this can be extracted from the ORACLE_HOME environment variable if set. On Windows the HKEY_LOCAL_MACHINE\Software\Oracle Registry key stores information about each Oracle home For each Oracle home the incident responder should locate the server's start up parameter file. This will be found in the "database" directory on a Windows system or the "dbs" directory on a *nix system. Generally the filename is

"spfilesid.ora" where "sid" is the database service identifier. This file contains information about where log and trace files etc are written to:

- audit_ file_ dest

- background_ dump_ dest

- core_ dump_ dest

- db_ recovery_ file_ dest

- user_ dump_dest

- utl_file_dir

- control_files

- db_create_file_dest

- db_create_online_log_dest_n

- log_archive_dest

- log_archive_dest_n The incident responder should also be aware that what is listed in the start up file may not actually be what settings the Oracle server is actually using 4) Get the previously executed SQL [17].

- Get a copy of the most recently executed SQL. This can be retrieved from the V$SQL fixed view. On Oracle 10g the query should be: SQL> SELECT LAST_ACTIVE_TIME, PARSING_ USER_ID, SQL_TEXT FROM V$SQL ORDER BY LAST_ACTIVE_TIME ASC; This will list the SQL that was executed by who and when from the V$SQL fixed view.

- Next in line should be the audit log. Everything should be selected from this table for later consumption and analysis. SQL > SELECT * FROM AUD$;

5) Getting a list of users and roles [17].

The incident responder should get a complete listing of all users on the system.

SQL > SELECT USER#, NAME, ASTATUS, PASS WORD, CTIME, PTIME, LTIME FROM SYS.USER$ WHERE TYPE# = 1;

6) Getting a list of dropped tables [17].

In 10g, if a user has dropped any tables and they have not been purged from the recyclebin then a list of dropped tables should be present. This may indicate evidence of an attack:

SQL > SELECT U.NAME, R.ORIGINAL_NAME, R.OBJ#, R.DROPTIME, R.DROPSCN FROM SYS.RECYCLEBIN$ R, SYS.USER$ U WHERE R.OWNER# = U.USER#;

7) Getting information about PL/SQL objects [17].

The source of PL/SQL objects should be retrieved and analyzed. Much of the source is encrypted or "wrapped" to use the Oracle term. The incident responder should obtain an "unwrapper" to examine the clear text as an attacker can modify a PL/SQL object and re-encrypt it to hide their attack.

8) Finishing up [17].

Once all queries have been executed the spool file should be closed and sqlplus can be closed.

SQL > SPOOL OFF SQL > QUIT Disconnected from Oracle Database 10g Enterprise Edition Release 10.2.0.2.0—Production with the Partitioning, OLAP and Data Mining options C:\oracle\product\10.2.0\db_1\BIN>

Once disconnected from the server an md5 checksum should be made of the spool file and recorded with a witness present.

3.3. Block Diagram

The proposed process flow has been shown in **Figure 3**. This diagram summarizes the different methods explained above. We have considered a database server which has been tampered by an unauthorized user. To detect the tamper the forensic analyzer has to focus and collect information from the specified locations such as redo logs, data blocks, audit trails, live response, views, oracle recycle bin, and system change number. Different sql commands and tools are available to retrieve the information. The obtained information should be stored in a server and a comparative analysis should be performed on the basis of different kinds of users and their grant roles, role membership, object privileges, system privileges, authentication and authorizations of each and every user of the database. The analysis should be performed in the presence of an in charge and reliable authority of the organisation or the database. A graph should be produced to show the variations in the expected performance. The graph can be considered as the summarized output of the digital evidences collected for detecting the tampering in the database. Hence using this output forensic analysis of the database server can be performed which can be used to identify who, when and where tampered the data.

The locations defined in the proposed block diagram shown in **Figure 4** can be useful to get information about the system and its user and hence will give guidance to process the flow shown in **Figure 4**. Analyzing the flow may give us the desired information.

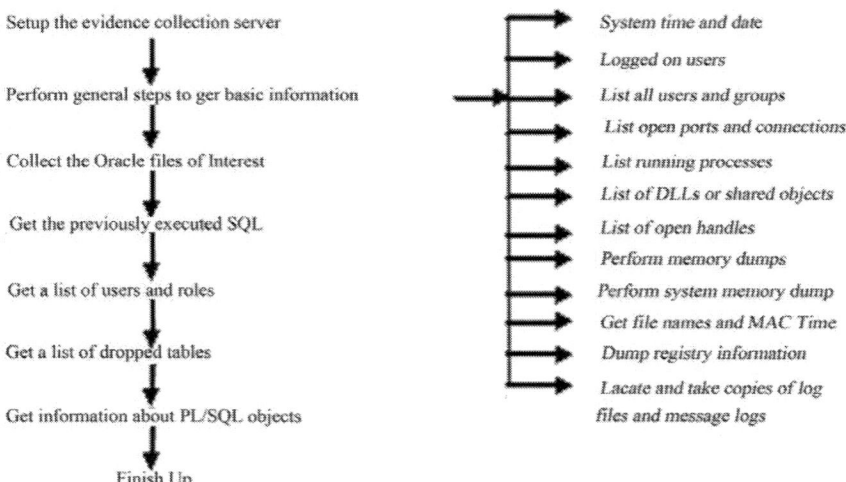

Figure 3. Process Flow.

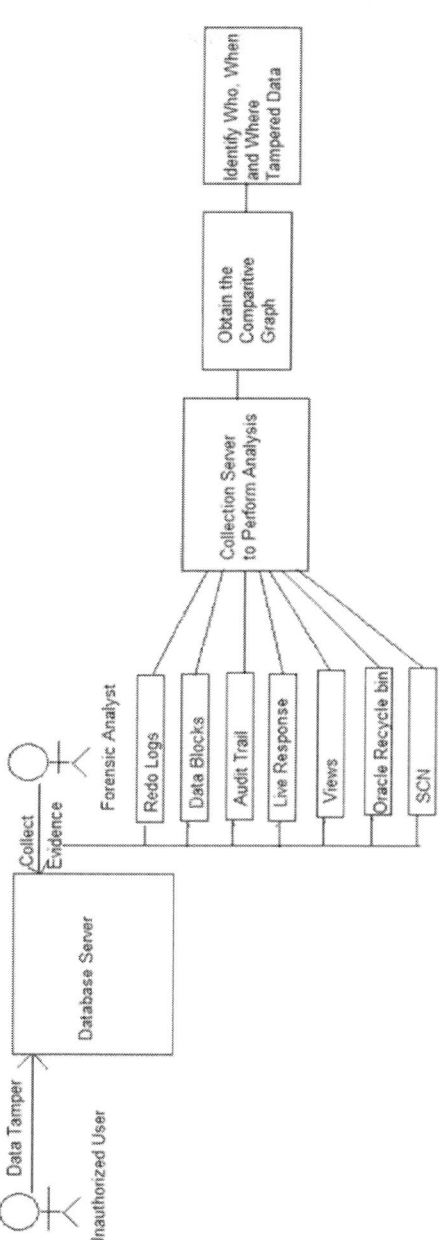

Figure 4. Proposed block diagram.

4. CONCLUSION

There are many ways of securing the database. The attackers have the methods to violate the security. Then comes the role of forensic analyst who should have a thorough knowledge of the basics of a database and also the information about the database on which he is going to perform the analysis. The forensic analyst should also be able to think from the attacker's point of view. Based on different cases, the digital evidences can be collected from the specified locations. If the intensions of the attacker are known identifying the attacked location may be easier. Thinking from the attacker's point of view this paper gives a contribution towards the identification of the general locations in a database for collecting the digital evidences.

REFERENCES

1. K. Loney and B. Bryla, "Oracle Database 10g DBA Handbook," McGraw-Hill, New York, 2005.

2. D. Litchfield, "Book on 'Oracle Forensics'," Wiley, Hoboken, 2008.

3. O. L. Carroll, S. K. Brannon and T. Song, "Computer Forensics: Digital Forensic Analysis Methodology," Computer, Vol. 56, 2008, pp. 1-8.

4. N. Aaron, "Oracle Database Security," ICTN 4040, Spring, 2006.

5. J. Azemović and D. Mušić, "Efficient Model for Detection Data and Data Scheme Tempering with Purpose of Valid Forensic

Analysis," Proceedings of the 2009 International Conference on Computer Engineering and Applications, Manila, 6-8 June 2009.

6. G. Miklau1 and D. Suciu, "Implementing a Tamper-Evident Database System," University of Massachusetts & University of Washington, Amherst & Washington DC, 2005.

7. J. Zhang, A. Chapman and K. LeFevre, "Do You KnowWhere Your Data's Been?—Tamper-Evident Database Provenance," Proceedings of the 6th VLDB Workshop on Secure Data Management, Lyon, 28 August 2009.

8. D. C. Lee, J. M. Choi and S. J. Lee, "Database Forensic Investigation Based on Table Relationship Analysis Techniques," Proceedings of the 2nd International Conference on Computer Science and Its Applications of the IEEE SCA, Jeju, 10-12 December 2009, pp. 1-5.

9. M. J. Malmgren, "An Infrastructure for Database Tamper Detection and Forensic Analysis," Bachelor's Thesis, University of Arizona, Tucson, 2007.

10. R. T. Snodgrass, S. S. Yao and C. Collberg, "Tamper Detection in Audit Logs," Proceedings of the 30th International Conference on Very Large Data Bases, Toronto, 31 August-3 September 2004.

11. "Oracle Forensics In a Nutshell 25/03/2007," 2007.

12. P. Finnigan, "Oracle Forensics," OUG Scotland, DBA SIG, 30 April 2008.

13. P. M. Wright, "Oracle Database Forensics Using LogMiner Option 3—Perform Forensic Tool Validation," Proceedings of the GCFA Assignment—GSEC, GCFW, and GCIH, London, 10 January 2005.

14. D. Litchfield, "Oracle Forensics Part 1: Dissecting the Redo Logs," NGSSoftware Insight Security Research (NISR), Next Generation Security Software Ltd., Sutton, 2007.

15. D. Litchfield, "Oracle Forensics Part 2: Locating Dropped Objects," NGSSoftware Insight Security Research (NISR), Next Generation Security Software Ltd., Sutton, 2007.

16. D. Litchfield, "Oracle Forensics Part 3: Isolating Evidence of Attacks against the Authentication Mechanism," NGSSoftware Insight Security Research (NISR), Next Generation Security Software Ltd., Sutton, 2007.

17. D. Litchfield, "Oracle Forensics Part 4: Live Response," NGSSoftware Insight Security Research (NISR), Next Generation Security Software Ltd., Sutton, 2007.

18. D. Litchfield, "Oracle Forensics Part 5: Finding Evidence of Data Theft in the Absence of Auditing," NGSSoftware Insight Security Research (NISR), Next Generation Security Software Ltd., Sutton, 2007.

19. D. Litchfield "Oracle Forensics Part 6: Examining Undo Segments, Flashback and the Oracle Recycle Bin," NGSSoftware Insight Security Research (NISR), Next Generation Security Software Ltd., Sutton, 2007.

20. D. Litchfield, "Oracle Forensics Part 7: Using the Oracle System Change Number in Forensic Investigations," NGSSoftware Insight Security Research (NISR), Next Generation Security Software Ltd., Sutton, 2008.

INDEX